TELECOMMUNICATIONS ENGINEERING INSTALLATION PROJECT DESIGN PACKAGE

PROJECT CONTACTS
(Provide details of the Networking and Customer Project Team).

Project Contact List

Project Team	(Customer) Project Team
Project Manager: Email: Telephone:	Project Manager: Email: Telephone:
Project Engineer: Email: Telephone:	Project Engineer: Email: Telephone:
Design Engineer: Email: Telephone:	Design Engineer: Email: Telephone:

PROJECT SITE INFORMATION
Address and Contact Details
(Provide the site address, on-site contacts and remote access information).

Table 1   Site Contact Details

Site Name	
Site Address	
Map Reference	
Site Contact	Name: Title: Telephone: Facsimile: Email: Out of Hours:
Hours of Operation	
Remote Access	

BUDGETING & FINANCES

PROJECT PACKAGE CHANGES & REVISIONS

TAB A LIST OF MATERIAL (LOM)

ITEM NO.	SCHEDULE OF SUPPLIES/ SERVICES	QUANTITY	UNIT	UNIT PRICE	AMOUNT

TAB B PROJECT INSTRUCTIONS

The project engineer/team leader prepares Tab B in sufficient detail to permit the installation and testing of the project without further clarification. This includes providing project drawings, sketches, maps, and circuit diagrams to show complete details of the installation.

Design Acceptance

The signatories below confirm that the design meets the requirements specified. The design is subject to change during or following staging.

By:_____   By:_____

Name: Name:

Title: Title:

Date:_____   Date:_____

TAB B Table of Contents

PROJECT SCOPE PURPOSE

STATEMENT STANDARDS

DOCUMENTS

REQUIREMENTS

TELECOMMUNICATIONS ROOMS MDF/IDF

ROOM DESIGN

CONSTRUCTION & DEMOLITION

ELECTRICAL REQUIREMENTS

GROUNDING & BONDING

HVAC SYSTEMS

INFRASTRUCTURE CABLING

MARKING & LABELING

CUSTOMER PREMISE EQUIPMENT (CPE)

FIBER OPTIC TESTS

NETWORK STAGING PLAN

ELECTRONICS EQUIPMENT SAFETY

NETWORK TOPOLOGY DIAGRAMS & EQUIPMENT DETAILS

NETWORK TESTING

NETWORK DOCUMENTATION

PHOTOGRAPHS

PROJECT DRAWINGS

ATTACHMENTS

PERFORMANCE WORK STATEMENT

FOR

UPGRADE NETWORK ELECTRONICS AND TELECOMMUNICATIONS

INFRASTRUCTURE FOR MANAGEMENT REGIONAL CAMPUSES

AT

ORGANIZATION TECHNOLOGY-FEE PROJECTS:

1011-0048/1213-34

3/19/13 Revision 0.0

1.0. PROJECT SCOPE

1.0.1. General Requirements. This Performance Work Statement (PWS) provides a comprehensive description of a new infrastructure and networking telecommunications upgrade project in support of the Technology-Fee program for MANAGEMENT Regional Campuses at Daytona. The MANAGEMENT CS&T office will have the option to perform all work in-house or contract the portion of this project for demolition and installation of telecommunications infrastructure within facilities and through ground trenching and/or preexisting manhole duct systems. The awarded contractor shall provide all personnel, equipment, tools, materials, vehicles, supervision and other items and services unless specified in this contract as MANAGEMENT furnished property. Work under this project will support the operations of MANAGEMENT Regional Campus at Daytona by providing a new highly reliable telecommunications network and wireless systems. The contractor or subcontractors shall also perform construction services as required to ensure the HVAC systems and electrical power requirements are available 24-hours per day, seven days per week. The contractor shall install all equipment and systems in accordance with original equipment manufacturers' technical manuals, MANAGEMENT specifications, policy, federal, state, local laws, and regulations. The contractor shall include mission-critical manufactures warranty documentation and applicable references.

1.0.2. Accident/Incident Reporting and Investigation. The contractor shall record and report all available facts relating to each instance of accidental property damage or personnel injury to the MANAGEMENT CS&T within two hours of the incident. The contractor shall secure the scene of an accident and wreckage until released by the accident investigative authority through the MANAGEMENT officials. If the MANAGEMENT officials elect to conduct an investigation the contractor shall cooperate fully until the investigation is completed.

1.0.3. Sub-Contractor Compliance. The contractor shall be responsible for ensuring sub-contractors satisfy the requirements set forth in the contract. The contractor shall include a provision in all subcontracts to require sub-contractor compliance with the terms and conditions of this contract.

1.0.4. Facilities Work Clearance Requests. The contractor shall prepare and coordinate for facilities work clearance requests prior to any trenching, digging, boring, construction, and demolition. The contractor shall prepare and coordinate and receive approval prior to any work requiring facility modification. The contractor shall coordinate with MANAGEMENT agencies to ensure appropriate markings are placed over existing infrastructure. The contractor can be held responsible for costs associated with the repair of any damages.

1.0.5. Personnel. The contractor is solely responsible for ensuring sufficient personnel are assigned to this contract and all personnel is qualified and certified to perform the requirements listed herein, including those qualifications and certifications required by the MANAGEMENT CS&T requirements to install or operate the equipment covered by this contract. The contractor shall ensure personnel also meet the criterion stated in their proposal to meet the requirements of this project.

1.0.6. Contractors shall obtain access passes used during the implementation period of this project. Prior to the issuing of access passes all contractor personnel will go through a background check to ensure there are no disqualifying factors. Upon contract award, the necessary applications for passes can be obtained by contacting MANAGEMENT CS&T.

1.0.7. Contractor identification cards may be issued for the performance period of the contract only. Contractor personnel will be made aware of the necessity for safeguarding identification cards issued and the requirement for reporting any identification cards lost.

1.0.8. If an employee is dismissed from employment, resigns, or if there is no longer a contractual requirement for the employee to enter MANAGEMENT facilities, the contractor shall ensure that the employee identification cards and any other identification, i.e., vehicle passes, issued to the employee are expeditiously returned to MANAGEMENT CS&T prior to final invoicing and payment. If all identification is not provided, a portion of the final payment (to be determined by the MANAGEMENT CS&T) will be withheld pending its submission.

1.0.9. All contactor personnel will comply with all MANAGEMENT emergency protection measures in the event there is an incident or threat to the campus.

1.0.10. The contractor shall provide contact information on key personnel to MANAGEMENT CS&T prior to commencing work and shall be responsible to update information on a continual basis as changes occur to ensure it is always current and correct throughout the life of the contract. All personnel shall be proficient in understanding, reading, writing, and speaking the English language. The key personnel list shall list the employees' names, a function they support, and whether they are supporting the work orders and what percentage of each.

1.0.11. Test Equipment. The contractor is responsible to have all necessary tests and support equipment on-site to support the installation of the project's cable infrastructure and network electronics.

1.0.12. Design Review Meetings. The contractor shall review program or project drawings and provide comments concerning communications requirements within ten working days from receipt when requested by MANAGEMENT CS&T. The contractor shall participate in designated design review meetings identified by MANAGEMENT CS&T.

1.0.13. Documents and Drawings. The contractor shall update and submit project documents and drawings to reflect as-built work within two weeks after completion of the associated task(s) unless specified differently within this section. The contractor shall produce, update, and post changes to drawings, plant records or documents. All documents and drawings established by the contractor are MANAGEMENT property and shall remain at the site and turned over to the CS&T for disposition upon contract completion. Failure to do so may result in withholding of final payment until all records are recovered or in a reduction in final payment for records lost or misplaced.

1.0.14. Contractor-Furnished Hazardous Material. The contractor shall develop this record to document bringing or using hazardous material on MANAGEMENT facilities. The contractor shall develop this record not later than 30 calendar days after the contract award and shall develop and maintain electronic updates to records quarterly thereafter. If no hazardous material is used in the performance of the contract during the quarterly period the contractor shall indicate "none" on the applicable quarterly record.

1.0.15. Contractor's Quality Control Program. The contractor shall establish, operate, and maintain a highly effective quality control program for all services provided under this project. MANAGEMENT CS&T will rely upon contractor documentation for all services performed subject to periodic review and validation. The contractor shall retain Quality Control documentation on-site in an orderly manner and make it available for MANAGEMENT CS&T review. The contractor shall provide quality assurance support for the entire life of the project. This support shall assist MANAGEMENT CS&T representative in performing random spot checks and installation systems acceptance tests. They shall be responsible for identifying system and telecommunications plant deficiencies and/or discrepancies throughout the life of the project and of final acceptance.

1.0.16. Security. The contractor shall follow MANAGEMENT security guidelines and at the end of each work period all MANAGEMENT
facilities, equipment, and materials shall be secured.

1.0.17. The contractor shall determine any potential problems with the projected work in areas containing asbestos, hazardous materials, and hazardous wastes with local procedures to be followed in the event such problems are encountered. Determine local procedures to be followed concerning the use, disposal, or reporting of use of hazardous materials. Material Safety Data Sheets (MSDS) shall be submitted "Hazardous Material Identification and Material Safety Data," for any hazardous materials (paints, solvents, cleaners, encapsulating compound, etc.) to be used in the performance of the contract. The MSDS must be on file with the MANAGEMENT Campus environmental management office prior to any use of the hazardous material.

1.0.18. Trenching, excavation, confined spaces entry, confined spaces atmospheric testing/forced air ventilation, and marking and barricading of open trenches are to be performed in accordance with Occupational Safety and Health Act (OSHA) standards.

1.0.19. Overtime. Overtime premiums are not reimbursed by the MANAGEMENT under the terms of the contract as it is included within the prices proposed; therefore changes in a shift or shift premiums/ reimbursement shall be negotiated solely between the contractor and its employees in accordance to applicable labor laws.

1.0.20. Corrective Action Report. When the Contractor fails to meet contract performance requirements; MANAGEMENT CS&T may prepare a corrective action report. This report is forwarded for review and subsequently processed through the contractor for action/explanation. Upon receipt, MANAGEMENT CS&T will review the response and determine whether to accept the contractor's position, determine corrective action (reduction in the monthly payment due to unacceptable performance, cure notice, show cause, etc.), or take other appropriate contractual action. MANAGEMENT CS&T reserves the right to make a partial payment for services performed prior to receipt and evaluation of the contractor's response to a corrective action report.

1.0.21. The contractor is also responsible for the removal and disposal of all abandoned cables and associated hardware in existing Facilities/Manhole/Duct System/ formerly owned by MANAGEMENT Regional Campus at Daytona Campus. All residues from this project shall be disposed of in accordance with MANAGEMENT, federal, and state environmental regulations.

1.0.22. Building Penetrations. All wall penetrations including inside buildings shall be restored to meet the required industry and MANAGEMENT fire ratings codes.

2.0. **PURPOSE STATEMENT**

2.0.1. Upgrade Network Electronics and Telecommunications Infrastructure for MANAGEMENT Regional Campuses at Daytona; Technology-Fee projects 1011-0048/1213-34. This project will consist of installing a new telecommunications infrastructure system and upgrading network electronics as part of the Technology-Fee program in support of MANAGEMENT Regional Campuses at Daytona.

2.0.2. This project will completely upgrade the telecommunications systems for the MANAGEMENT Daytona campus with the intent of providing greater network reliability, redundancy, and performance.

2.0.2. The 1st phase of the project will require designation, demolition, and construction of new telecommunications rooms in buildings 140 and 150. Each building will require one Main Distribution Frame (MDF) located on the first floor and Intermediate Distribution Frame (IDF) on every other floor. The other option will allow for the design of only one designated MDF (BLDG 150, Room 105B) and all other telecommunication rooms designated as IDF's. The proposed HVAC work will be integrated with the Organization (DSC) water chiller system with each MDF and IDF receiving new air handlers to be installed in accordance with MANAGEMENT CS&T guidelines and industry standards. The heat load capacities should fall in the range of approximately 30,000 BTU for MDF and 10,000 to 15,000 BTU ranges for IDF rooms. All HVAC equipment and water piping shall maintain appropriate distances from networking electronics and equipment. Each telecommunications room will require the contractor to remove existing flooring, prepping walls, core boring between floors, and electrical roughing and finishing. MANAGEMENT CS&T shall be responsible for mounting ¾ inch fire rated plywood is installed in all MDF and IDF rooms. All framing and associated hardware to support the installation of cables and equipment will be installed in accordance with industry and MANAGEMENT CS&T Design Guidelines. The contractor shall be responsible for the removal and disposal of all old cabling and associated hardware from facilities after project completion. MANAGEMENT CS&T will remove all electronics and dispose of them in accordance with MANAGEMENT policies.

2.0.3. The 2nd phase will include scheduled cutover and the installation of new cabling infrastructure consisting of Cat 5E and single-mode fiber optic cables between MDF and all other IDF rooms within buildings 140 and 150. The Bright House Metro Ethernet fiber optic cables along with electronics can be moved from the existing rack in building 150, room 118 can be relocated to the MDF located in building 140, room 105B. If relocation of Bright House single-mode fiber optic cable cannot be relocated due to cost, the existing switch in building 150 room 118 shall be coordinated with Bright House and relocated in MDF in building 140, room 105B. Originating from the MDF in building 140, room 105B the installation team shall run a total of eight (8) each 24-strand single-mode fiber optic cables in home run fashion to all IDF's including room 118 and DSC library facilities. Category 5E compliant plenum-rated station cable and all associated hardware will be installed to all locations as specified in the project plan and drawings. Existing raceway and J-hooks shall be reused if they are in compliance with MANAGEMENT CS&T specifications and guidelines. During the pre-implementation survey, the installation team or contractor shall determine the condition of the existing raceway and J-hooks located at above false ceilings located throughout both facilities.

2.0.4. The 3rd phase of this project will require the installation of a new core router and switching equipment at each MDF and IDF in buildings 140 and 150. Wireless access points will be installed throughout both buildings at specified locations in accordance with the radio frequency survey. Installation of VOIP telephones will be the responsibility of the DSC telecommunications office, MANAGEMENT will provide assistance with configuration and assignment of the port on the newly installed switching equipment. The new router will provide wireless management services through VPN configuration across the Bright House Metro Ethernet WAN connection to the MANAGEMENT main campus network.

2.0.5. The 4th phase of this project will require complete testing and documentation consisting of updating drawings and records. The contractor shall be responsible for the removal and disposal of all abandoned cables and associated hardware. The final step of the project will consist of a quality assurance inspection and documentation. MANAGEMENT CS&T can conduct a customer survey in order to evaluate system performance and quality of work.

3.0. STANDARDS DOCUMENTS

3.0.1 All telecommunications installation tasks shall be performed in accordance with the most recent industry standards and internal MANAGEMENT CS&T design guidelines. Any and all issues with standard installation practices shall follow the chain of command protocol for approval prior to the implementation of work with MANAGEMENT CS&T. The following are the main publications and standards used to perform this project and all other listed standards to provide a larger scope of publications available:

Publication Number	Date	Title
MANAGEMENT CS&T) Design Guidelines:	7-1-11	MANAGEMENT Computer Services & Telecommunications (MANAGEMENT
NFPA 70	2011	National Electrical Code
ANSI/TIA/EIA-568-C.1		Commercial Building Telecommunications Cabling Standard
ANSI/TIA/EIA-569-B		Commercial Building Standard for Telecommunications Pathways and Spaces
ANSI/TIA/EIA-606-A		Administration Standard for the Telecommunications The infrastructure of Commercial Buildings
ANSI/J-STD-607		Commercial Building Grounding and Bonding Requirements for telecommunications
BICSI		BICSI Telecommunications Distribution Methods Manual (TDMM) 11th Edition
ANSI/EIA-310-D		Cabinets, Racks, Panels, and Associated Equipment
TIA-568-B		Commercial Building Telecommunications (568B-1, 568B-2, 568B-3) Cabling Standard
BICSI TDM Manual 11th Edition		Building Industries Consulting Services International Telecommunications Distribution Methods (TDM) Manual
BICSI OPDRM		Outside Plant Design Reference Manual

FEDERAL SPECIFICATIONS AND STANDARDS
The following safety references (Code of Federal Regulations (CFR)) are not intended to exclude any other references but are intended to identify those safety references, which are most common to the efforts required by the PWS. The Contractor is required to abide by any and all applicable Federal, State, and local codes and safety requirements.

Publication Number	Date	Title

Publication Number	Date	Title
CFR Title 47 Part 68		Telecommunications Federal Communications Commission (FCC) Connection of Terminal Equipment to the Telephone Network
7 CFR 1755.370		Rural Utility Service (RUS) Specification for Seven Wire Galvanized Steel Strand
EPA CFR 40 Parts 260, 261, 262, 263, 264, 265		Hazardous Waste Management System

(Copies of the above documents may be obtained from the GPO Regional Printing Procurement Office, 928 Jaymore Rd, Suite A-190, Southampton, PA 18966-3820, Phone (215) 364-6465; FAX (215) 364-6479 or http://www.gpoaccess.gov/cfr/retrieve.html.)

Publication Number	Date	Title
OSHA CFR 29 Part 1910		Occupational Safety and Health Standards
OSHA CFR 29 Part 1910.268		Telecommunications
OSHA CFR 29 Part 1910.1101		Asbestos
OSHA CFR 29 Part 1926.650, 651, and 652		Excavation
OSHA CFR 29 Part 1910.120		Hazardous Waste Operation and Emergency Response
OSHA CFR 29 1910.146		Permit-required Confined Spaces
OSHA CFR 29 Part 1910.147		The Control of Hazardous Energy
OSHA CFR 29 Part 1910.1200		Hazard Communications (Lockout/Tagout)

(Occupational Safety and Health Administration (OSHA) documents may be obtained at www.osha.gov.)

NATIONAL STANDARDS

Publication Number	Date	Title
NFPA 70-08		National Electrical Code (NEC)

(Copies of the above document may be obtained from the National Fire Protection Association (NFPA), 11 Tracy Drive, Avon, MA 02322, Phone 1-800-344-3555, Fax 1-800-593-6372.)

Publication Number	Date	Title
Telecommunications Industry Association (TIA)/Electronic Industry Alliance (EIA)-232-E Data Interchange		The interface between Data Terminating Equipment and Employing Serial Binary

(Copies of the above document may be obtained from the Global Engineering Document, 15 Inverness Way East, Englewood, CO 80112, Phone 1-800-854-7179.)

Publication Number	Date	Title
ANSI-ICEA-83-596		Fiber Optic Premises Distribution Cable

(Copies of the ICEA standard may be obtained Global Engineering Documents, 15 Inverness Way East, Englewood CO 80112; Tel: (800) 854-7179; FAX: (303) 397-2740; or Email: global@ihs.com Website: http://global.ihs.com)

Publication Number	Date	Title
ANSI C2-1997		National Electrical Safety Code (NESC)
ANSI/TIA/EIA -470-B		Telephone Instruments with Loop Signaling
ANSI/TIA/EIA-568-B.1/B.2		Commercial Building Telecommunications Cabling Standard—Part 1: General Requirements & Part 2: Balanced Twisted Pair Cabling Components
ANSI/TIA/EIA-568-B.3		Optical Fiber Cabling Components Standard

Publication Number	Date	Title
ANSI/TIA/EIA-569-A		Commercial Building Standard for Pathway and Spaces

(Copies of the above ANSI/TIA/EIA documents may be obtained from the American National Standards Institute, 25 W 43rd Street, 4th Floor, New York City, NY 10036, Phone (212) 642-4900, Fax (212) 398-0023.)

Publication Number	Date	Title
National Electronic Manufacturers' Association (NEMA) TC-2		Electrical Polyvinyl Chloride Conduit

(Copies of the NEMA documents may be obtained from National Electrical Manufacturers' Association, 1300 N. 17th Street, Ste 1847, Rosslyn, VA 22209, Phone (703) 841-3200, Fax (703) 841-3300 or at www.nema.org)

OTHER GOVERNMENT DOCUMENTS

Publication Number	Date	Title
REA TE&CM Section 451.2		Shield Continuity
REA TE & CM Section 823		Electrical Protection by Use of Gas Tube Arresters
REA TE & CM, Sect 825 Issue No. 3		Situations Requiring Special Protection
REA Specification PE-33		Shield Bonding Connectors
REA Specification PE-52 Bulletin 345-54		Telephone Cable Splicing Connectors
REA Specification PE-74 Bulletin 345-72		REA Specification for Filled Splice Closures
REA Specification PE-80 Bulletin 345-83		REA Specification for Gas Tube Surge Arresters
RUS 1751F-630		Design of Aerial Plant
RUS 1751F-635		Aerial Plant Construction
RUS 1751F-640		Design of Buried Plant-Physical Considerations
RUS 1751F-641		Construction of Buried Plant
RUS 1751F-642		Construction Route Planning of Buried Plant
RUS 1751F-643		Underground Plant Design
RUS 1751F-644		Underground Plant Construction
RUS 1751F-650		Aerial Plant Guying and Anchoring
RUS 1751F-805		Electrical Protection at Customer Locations
RUS 1751F-815		Electrical Protection of Outside Plant
RUS 1753F-150		Specifications and Drawings for Construction of Direct Buried Plant
RUS 1753F-151		Specifications and Drawings for Construction of Underground Plant
RUS 1753F-152		Specifications and Drawings for Construction of Aerial Plant
RUS 1753F-153		Specifications and Drawings for Service Installation at Customer Access Locations
RUS 1753F-205 (PE-39)		REA Specification for Filled Telephone Cable
RUS 1753F-207 (PE-87)		REA Specification for Terminating Cables
RUS 1753F-208 (PE-89)		REA Specification for Filled Telephone Cables With Expanded Insulation
RUS 1753F-302 (PE-91)		RUS Specification for Outside Plant Housings and Serving Area Interface Systems
RUS 1753F-401 (PC-2)		RUS Standard for Splicing Copper and Fiber Optic Cables
RUS 1753F-601 (PE-90)		REA Specification for Filled Fiber Optic Cables

Publication Number	Date	Title
REA Standard (PC-5A) RUS 1753F-153		Specifications and Drawings for Service Installation at Customer Access Locations

(Copies of the REA and RUS documents may be obtained from the Telephone Operations and Standard Division, Rural Electrification Administration, U.S. Department of Agriculture, Washington D.C., 20250, Phone (202) 720-8674.)

4.0. REQUIREMENTS

4.0.1. All electronic and information technology procured through this project and any resulting contract must meet the applicable accessibility standards unless an agency exception to this requirement exists. The MANAGEMENT CS&T group has a requirement to support and maintain the MANAGEMENT Regional Campus Daytona telecommunications infrastructure, network electronics, and wireless systems. Contract services shall be utilized to support CS&T in demolition and construction of existing and new telecommunications rooms. Additionally, contract services can be used to provide HVAC and electrical systems on as needed bases.

4.0.2. MANAGEMENT CS & T shall procure networking equipment, Category 5E cabling, fiber optic cabling, connecting materials, wireless access equipment, and all associated auxiliary infrastructure materials necessary for the complete installation of a fully functioning wired and wireless network. The MANAGEMENT CS&T office will have the option to perform all work in-house or contract the inlay of fiber cabling between buildings and telecommunications rooms, through ground trenching and/or preexisting manhole duct systems. The contractor shall terminate the fiber optic cabling and Cat 5E cabling and perform pre and post-testing between the telecommunications rooms in buildings 140 and 150 to include the Bright House fiber optic switch and DSC library. All networking electronics will be installed and configured in-house by MANAGEMENT CS&T technicians.

4.0.3. All fiber optic cables shall be 24-strand single-mode or higher and Category 5E compliant plenum-rated 4-pair 24-AWG cable brands shall meet industry specifications and MANAGEMENT CS&T Design Guidelines: 7-1-11, Revision 9.3.

4.0.4. Contractors need to consult MANAGEMENT CS&T for clarification of current media types and standards before completing telecom designs, purchasing materials, and commencing work. Basic requirements for all new building structures will include underground service entrance ducts, telecommunication rooms, a conduit riser system, anda a horizontal cable distribution system as defined in MANAGEMENT CS&T Design Guidelines: 7-1-11, Revision 9.3.

4.0.5. The contractor will be required to purchase, install, test, and document all communications wiring as specified by CS&T. The contractor may be required to run conduits and core drills within the scope of selected projects. The contractor shall be required to submit all materials for review and approval prior to any procurement.

4.0.6. Telecommunications networking electronic equipment will normally be provided (purchased and installed) by the MANAGEMENT CS&T technicians, but in some instances,s contractors can be specified for network configuration, design, consulting, wireless radio frequency surveys, and any other requirements on as needed bases.

4.0.7. Cable Tray, Raceway System, and J-Hooks: Overhead cable tray systems for data transmission and conduit shall be provided as a telecommunication distribution system throughout the facility. The cable tray system shall be an open steel mesh type, also known as a wire basket tray system designed for ease of access. The cable tray system shall be "UL Classified" as suitable for use as an electrical conductor for grounding and bonding. Where practical, provide home run conduits from outlets to frame rooms, as a primary choice of the raceway. MANAGEMENT CS&T Design Guidelines: 7-1-11, Revision 9.3.

5.0. TELECOMMUNICATIONS ROOMS MDF/IDF
(Building 140, Room 105B MDF, 208A IDF, 310B IDF – Building 150, Room 103 MDF/IDF, Room 200E IDF, Room 300E IDF, 400F IDF)

5.0.1. Telecommunications rooms shall be of the minimum square footage of 100 square feet in size and shall have both an HVAC supply and return vent. For specific power requirements for the primary distribution, room refers to drawing templates and MANAGEMENT specifications and standards. MANAGEMENT CS&T Design Guidelines: 7-1-11, Revision 9.3.

5.0.2. All buildings will have an MDF (Main Distribution Frame), generally located on the lowest level of the building. In this project, it is recommended to only have one MDF located in building 140, room 105B. This room will be a walk-in room with recommended dimensions of 10-feet deep x 10-feet wide (double doors may be required as directed by CS&T) unless otherwise specified and shall be separated from other electrical, mechanical, and housekeeping spaces. This project will utilize existing doors and possibly upgrade the lock system to the main campus door key entry card system.

5.0.3. IMPORTANT: Telecommunications rooms shall be located in the building such that the maximum cable distance from a communications outlet to the termination point in a room does not exceed 90 meters. In addition, 10 meters is allowed for interconnecting cable in the room and for station cable in the office for a total cable distance of 100 meters. Concerning maintenance loop and slack requirements for fiber optic cables follow industry and MANAGEMENT CS&T standards. IDF rooms will have minimum inside dimensions of 10' deep x 10' wide minimum standards unless otherwise specified by CS&T. Double doors may be specified and remain an option.

6.0. ROOM DESIGN
(Building 140, Room 105B MDF, 208A IDF, 310B IDF – Building 150, Room 103 MDF/IDF, Room 200E IDF, Room 300E IDF, 400F IDF). The telecommunications rooms will be minimally equipped with the following:

6.0.1. Ceilings shall be open to structure in order to accommodate fire rated plywood walls and new lighting fixtures and equipment framing and associated hardware.

6.0.2 Room lighting (50-foot candles measured at 3-feet AFF) with a motion sensor and a light switch at the entrance of the telecommunications room.

6.0.3. All rooms shall be located away from any source of water damage. No water-carrying pipes shall be permitted run through or within the ceiling space or floor of MDF/IDF rooms, exception with pipes associated with any required fire protection system. No showers, toilets, or similar wet rooms/areas shall be adjacent to or above telecommunications rooms. In addition, as much as practicable rooms shall be located away from electrical transformers, generators, air conditioning units, and or radio transmission equipment.

6.0.4. Door locks keyed to Telecommunications room standards. The intent is to have an electronic door access device on all telecom room doors. MANAGEMENT main campus uses the card system and minimally measures should be taken to ensure that door frames are equipped to accommodate these devices.

6.0.5. Doors will be one-hour fire-rated and designed to provide full access to the room. These doors will be designed as either single or double doors. All doors will open into corridors or common space. The intent of this requirement is to ensure that all telecommunications rooms are accessible without having to access other spaces.

6.0.6. Doors, frames, hardware, and locks are not being included in this project for funding purposes. A customer has the option to include these additional requirements during and after project completion as funds become available. Electronic Locks on both the outer suite and the IDF/MDF room will be used if funding permits.

6.0.7. All existing locks in building 140 and 150 are beginning to fail. A replacement electronic lock system for the failing locks is an unfunded project at this time.

6.0.8. MANAGEMENT Regional Campuses will follow up with Presidio and DSC to see if we can standardize locks for doors needing an electronic entry in the buildings, so whatever is chosen for the IDF/MDF rooms can be used elsewhere in the facility

6.0.9. Standard 8-feet x ¾-inch fire-retardant plywood, grade A-C equivalent, fully lining all walls to a height of eight-feet and beginning at a minimum of 6-inches off floor level. Plywood shall be painted with a minimum of two coats white or grey fire-resistant paint. At least one fire-retardant stamp shall be left exposed on each sheet of plywood. The fire-rated plywood shall be painted white on the side facing the window in order to match the décor of the facilities. The contractor shall frame walls in such to maintain access by framing around window and allowing for easy removal of fire rated plywood. The contractor shall not paint the windows and leave them as is in order to maintain facilities décor.

6.0.10. A 12-inch wide by 1-1/2-inch deep tubular steel ladder runway installed around the perimeter of room located at 4-inch off wall surface and approximately 7-feet above the finished floor. Equipment ladder runway with bottom drops out devices at floor racks as required ensuring cable bundling is fully supported to maintain proper bend radius.

6.0.12 Air-conditioned air which meets and typically exceeds normal building standards for office space as a minimum. Heat load requirements shall be calculated based upon electronic network switch equipment that will be installed in each room. Conditioned air shall be independently controlled for each telecom room and provided 24-hours/7-days a week, 365-days per year. The HVAC design for all communications rooms will need to make certain that air is exchanged out of the room. CS&T will review and approve all HVAC designs including mean temperature and number of air exchanges per hour.

6.0.13. Design efforts shall attempt to locate telecommunications room cooling equipment such that it is not located in the telecom room ceiling space. This project proposes HVAC equipment located above the entrance doorway.

6.0.14. Signage consistent with MANAGEMENT signs standards labeling room as "Telecommunications Room." Installer and/or contractor need to coordinate with CS&T standards for details concerning sign standards.

6.0.15. High rise buildings, greater than one floor (stacked closets) must have a 1/8" vertical strength member for strain relief of riser cables in each riser sleeve. This strength member is not part of cable but is a separate braided cable on which riser cables are supported on. Contractor to submit product cut sheets prior to beginning work.

6.0.16. All equipment racks and associated hardware types and brands are specified in the MANAGEMENT CS&T Design

Guidelines: 7-1-11, Revision 9.3. Minimum equipment rack specifications require standard 19-inch wide x 7-feet floor mounted frame. Vertical and horizontal cable management hardware should be used in order to provide better cabling management practices. Floor mounted equipment racks shall also be secured at their top with horizontal ladder racking to prevent swaying of the equipment racks.

6.0.17. Equipment Racks Containing Category 5E 24-port, 48-port, and higher patch panels shall be by brand name and in accordance with MANAGEMENT CS&T Design Guidelines.

6.0.18. Equipment Racks containing network electronics shall have a flush mount horizontal power strip equipped with 5-20R electrical receptacles.in accordance with MANAGEMENT CS&T Design Guidelines: 7-1-11, Revision 9.3.

7.0. CONSTRCUTION & DEMOLITION

7.0.1. Demolition will be required in Daytona Campus MDF and IDF rooms consisting of framing, drywall, painting, and flooring per MANAGEMENT CS&T specs. All windows in the MDF and IDF rooms shall not be painted and framed to include drywall and finished with primer paint. The fire-rated plywood shall be painted white on the side facing the window in order to match the décor of the facilities. The contractor shall frame walls in such to maintain access by framing around window and allowing for easy removal of fire rated plywood. The contractor shall not paint the windows and leave them as is in order to maintain facilities décor. The flooring shall consist of VCT with vinyl base; floor coloring should be determined by customer and MANAGEMENT CS& T specifications. Floors shall be covered with static resistant VCT. Sealed concrete will not be permitted.

7.0.2. The contractor shall obtain all work permits from appropriate facilities manager prior to starting any work requiring digging and/or core boring in facilities throughout the project and campus. It is estimated a total of 10-total core 4½-inch diameter bores, sleeved and properly grounded as required: Building 140: Room 105B (2-Each) 4¼-inch core bores on the ceiling and one 4-inch entrance on an exterior wall. Room 204D (2- Each) 4¼-inch core bores on the ceiling. Building 150: Room 103 (2-Each) 4¼-inch core bores on the ceiling. Room 200E (2-Each) 4 ¼ inch core bores on the ceiling. Room 300E (2-Each) 4¼-inch core bores on the ceiling.

7.0.3. Telecommunications rooms will be connected by a minimum of two 4 1/4-inch sleeves or conduits. Both sleeves and conduit will be located along the rear wall, in the left rear of the room and will be stubbed 4 inches above the finished floor, 4-inches off the rear wall. Sleeves shall be fitted with 4-inch set-screw or compression connectors and screw-on type plastic bushings. Bushings shall be installed before any cable is pulled through riser conduits/sleeves. If riser conduit extends between rooms that are not stacked, a marked pull tape shall be provided. There shall be no more than two 90 degree bends in the riser conduit runs between rooms without installation of a pull box 24-inch x 24-inch x 8-inch deep. LB fittings will not be accepted.

7.0.4. The contractor shall install approximately 70-feet of two 4-inch schedules 40 conduits between building 14, MDF room 105B and building 150, room 118. Existing conduit on the exterior wall entering building 150, room 118 can be utilized to install fiber optic cables. Cap one 4-inch schedule 40 conduits and leave buried for future use. The contractor shall dig underneath the sidewalk without breaking existing concrete and place a minimum of 4-inch sand under and above conduits and backfill without stones greater than one inch in diameter.

8.0. ELECTRICAL REQUIREMENTS
(Building 140, Room 105B MDF, 208A IDF, 310B IDF – Building 150, Room 103 MDF/IDF, Room 200E IDF, Room 300E IDF, 400F IDF)

8.0.1. All electrical panels and devices shall be UL listed and installed per manufacturer's instructions and meet the requirements of the National Electric Code (NEC) standards.

8.0.2. All electrical work on this project consists of removing existing drop ceiling and lighting fixtures in the new MDF/IDF rooms and removing electrical systems from old telecom rooms. The installation will consist of installing the new electrical wiring as needed with each telecommunications room receiving a minimum of 3- dedicated 20-Amp, 120-Volt electrical circuits to be utilized by telecommunications equipment. One (1- Each) dedicated 30-Amp, 120-Volt circuits will be installed in every telecommunications room and to be utilized by telecommunications equipment. A single (1-Each) 20-Amp, the 120-Volt dedicated circuit will be installed in every telecommunications room and to be utilized for HVAC fan coil. All lighting shall include motion sensors and be secured on the ceiling and centrally located in each and every telecommunications
rooms as specified by MANAGEMENT CS&T specifications. All unused and old electrical cabling in

telecommunications rooms and above ceilings should be removed at the source from the main electrical panel.

8.0.3. According to MANAGEMENT CS&T Design Guidelines: 7-1-11, Revision 9.3. Each wall of each telecommunications room shall have a minimum of one dedicated quad 120 Volt AC, 20 Amp L5-20R electrical receptacles fed by emergency generator power. The electrical contractor shall coordinate with MANAGEMENT CS&T on the exact locations of electrical receptacles in each MDF/IDF.

8.0.4. Each MDF/IDF shall have a minimum: Two dedicated L630R electrical receptacles served by commercial power and two dedicated L1430R electrical receptacles served by emergency generator power. Those receptacles are typically installed on the wall behind equipment racks. The electrical contractor shall coordinate with MANAGEMENT CS&T PM, on the exact locations of electrical receptacles in the MDF/IDF. Electrical lights, electrical receptacles, card access systems, and HVAC units serving all MDF/IDF shall also be supplied via emergency generator power during power outages. MANAGEMENT CS&T Design Guidelines: 7-1-11, Revision 9.3.

8.0.5. Important: During the 1st stage of this project electrical contractor shall rough-in electrical wiring above MDF/IDF rooms and securely coil slack above the ceiling. An electrical contractor shall only energize and finish electrical work after the MDF/IDF rooms have been completely been framed and racked. The electrical contractor needs to coordinate finish work prior to the installation project manager for the exact location of electrical receptacles.

8.0.5. GROUNDING & BONDING

8.0.6. Grounding shall originate from the main facilities' grounding system to a grounding bus bar located in each MDF/IDF. (Building 140, Room 105B MDF, 208A IDF, 310B IDF – Building 150, Room 103 MDF/IDF, Room 200E IDF, Room 300E IDF, 400F IDF) as determined by the installation team and MANAGEMENT CS&T Design Guidelines: 7-1-11, Revision 9.3, specifications.

8.0.7. Bus bars shall be cleaned prior to fastening the conductors to the bus bars, and anti-oxidant grease shall be applied to the contact area in order to control corrosion and reduce contact resistance. MANAGEMENT CS&T Design Guidelines: 7-1-11, Revision 9.3.

9.0. HVAC SYSTEMS

9.0.1. The HVAC mechanical ductwork will be limited to all new MDF/IDF rooms included in this project (Building 140, Room 105B MDF, 208A IDF, 310B IDF – Building 150, Room 103 MDF/IDF, Room 200E IDF, Room 300E IDF, and 400F IDF). The project will require a total of seven (7) total chilled water fan coil units in each telecommunications room. The HVAC chilled water fan coil units will be approximately between 10,000 and 30,000 BTU or as determined by the equipment load requirements. This project will be utilizing the Extreme Networking Equipment; Summit X460-48P switches which have a heat dissipation specification of 223 Watt 760 BTU/HR per PSU. This calculation only accounts for one networking switch and will need to calculate any additional number of switches and other equipment placed in the telecommunications rooms to get proper HVAC load requirements. In addition, each equipment rack will require an uninterrupted battery backup power supply in which need to be included in the overall total heat dissipation calculations.

9.0.2. No HVAC equipment or any associated chilled water, refrigerant, or condensate water piping shall be installed in MDF/IDF rooms. HVAC equipment and piping to serve MDF/IDF's shall be installed outside MDF/IDF's and only duct shall enter above door entries if possible. MANAGEMENT CS&T Design Guidelines: 7-1-1, Revision 9.3.

9.0.3. Telecom Rooms used as MDFs, typically contain MANAGEMENT provided network electronics generating between

15,000 and 30,000 BTUs per hour of heat load, and shall be cooled at a constant 68 degrees Fahrenheit. MANAGEMENT CS&T Design Guidelines: 7-1-11, Revision 9.3.

9.0.4. Telecom Rooms used as IDFs, typically contain MANAGEMENT provided network electronics generating at least

10,000 BTUs per hour of heat load, and are to be cooled at a constant 72 degrees Fahrenheit. MANAGEMENT CS&T Design Guidelines: 7-1-11, Revision 9.3.

10.0. INFRASTRUCTURE CABLING
(Building 140, Room 105B MDF, 208A IDF, 310B IDF – Building 150, Room 103 MDF/IDF, Room 200E IDF, Room 300E IDF, and 400F IDF).

10.0.1. Fiber optic riser cabling will connect the MDF to each individual IDF with a minimum (unless otherwise noted) 24-strand 9.0-micron single-mode fiber optic cabling, and 24-strand 62.5-micron multi-mode fiber optic cabling, if required in one, continues to pull. Fiber optic cabling will be terminated in fiber distribution centers mounted in floor equipment racks. See MANAGEMENT CS&T Design Guidelines: 7-1-11, Revision 9.3 for further details.

10.0.2. In building 140, room 105B will serve as the MDF and remain as the fiber backbone for connecting the MANAGEMENT Bright House Metro-Ethernet link in Building 150, room 118, Organization network, and Building 150. In addition, room 105B will serve as the location for the WAN router and core switch room linking buildings 140 and 150 with direct 24-strand single-mode fiber optic cables to each frame enclosure located on the individual floor in home-run fashion. In total there will be eight (8) 24-strand single-mode fiber optic cables that will be distributed to and within building 140 and 150 to include separate fiber cable runs to Bright House and DSC library facilities. A 192-port single-mode fiber optic patch panel will need to be rack mounted and installed in building 140, room 105B MDF. Also, MDF room 105B shall have 2-each 4
¼ inch core drilled bores on ceiling connecting the server room 204D on the 2nd floor. Fiber optic cable shall be installed through room 204D continuing to the new IDF in room 208A in building 140. Install the 2nd
24-strand single-mode fiber optic cable straight through to within building 140 to room 310IDF. Install five (5) 24-strand single-mode fiber optic cables in a single pull starting at the MDF in building 140, room 105B via newly installed schedule 40 conduits to building 150 room 118 and with one (1) 24-strand single-mode fiber optic cable with run into building 150, room 118 reserved for Bright House Metro Ethernet connection, the 2nd pull of 24-strand single-mode fiber optic cable will be run into the MDF/IDF room 103, the 3rd pull of 24-strand single-mode fiber optic cable will be run into room 200E, the 4th pull of 24-strand single-mode fiber optic cable will be run into room 300E, and the 5th pull of 24-strand single-mode fiber optic cable will be run into room 400F. All fiber optic cable maintenance loop should be installed in the MDF room in a wall-mounted enclosed standard wall mounted fiber optic panel.

10.0.3. Intermediate Distribution Frame (IDF) will be centrally located on each level of the building, and ideally arranged in a "stacked" fashion. Minimally, there shall be a room for every 10,000 square feet of office floor space. Rooms shall provide a minimum quantity of two (2) 4 ¼-inch core bores with 4-inch sleeves/conduits through the floor in the back left-hand corner or as specified by MANAGEMENT CS&T. Installer or contractor if required will install a minimum quantity of three (2) 1 ¼-inch plenum-rated inner-ducts in one of the four-inch backbones and/or riser conduits between floors.

10.0.4. In building 140, between MDF/IDF rooms 105B, 208A, and 310B provide a minimum quantity of two (2) 4 ¼-inch core bores with 4-inch sleeves/conduits thru the floor in a back left-hand corner or as specified by MANAGEMENT CS&T. In MDF room 105B install drill one 4-inch hole on the exterior wall and mount conduits and pull boxes on exterior wall entering MDF room 105B.

10.0.5. In building 140 in IDF rooms 208A install one (1) each 24-strand single-mode fiber optic and one (1) each 24-port single-mode fiber-optic patch panel rack mounted. Any and all fiber optic cable maintenance loop should be installed in the MDF room in a wall-mounted enclosed standard fiber optic panel.

10.0.6. In building 140 in IDF rooms 310BA install one (1) each 24-strand single-mode fiber optic and one (1) each 24-port single-mode fiber optic patch panel rack mounted. Any and all fiber optic cable maintenance loop should be installed in the MDF room in a wall-mounted enclosed standard fiber optic panel.

10.0.7. In building 150, between rooms 103, 200E, 300E, and 400F provide a minimum quantity of two (2) 4 ¼-inch core bores with 4-inch sleeves/conduits through the floor in a back left-hand corner or as specified by MANAGEMENT CS&T.

10.0.8. In building 150 in MDF/IDF rooms 103 installs one (1) each 24-strand single-mode fiber optic and one (1) each 24-port single-mode fiber-optic patch panel rack mounted. Any and all fiber optic cable maintenance loop should be installed in the MDF room in a wall-mounted enclosed standard fiber optic panel.

10.0.9. In building 150 in IDF rooms 200E install one (1) each 24-strand single-mode fiber optic and one (1) each 24-port single-mode fiber optic patch panel rack mounted. Any and all fiber optic cable maintenance loop should be installed in the MDF room in a wall-mounted enclosed standard fiber optic panel.

10.0.10. In building 150 in IDF rooms 300E install one (1) each 24-strand single-mode fiber optic and one (1) each 24-port single-mode fiber optic patch panel rack mounted. Any and all fiber optic cable maintenance loop should be installed in the MDF room in a wall-mounted enclosed standard fiber optic panel.

10.0.11. In building 150 in IDF rooms 400F install one (1) each 24-strand single-mode fiber optic and one (1) each 24-port single-mode fiber optic patch panel rack mounted. Any and all fiber optic cable maintenance loop should be installed in the MDF room in a wall-mounted enclosed standard fiber optic panel.

10.0.12. In building 150 in Bright House Metro Ethernet rooms 118 installs one (1) each 24-strand single-mode fiber optic and one (1) each 24-port single-mode fiber-optic patch panel wall mounted. Any and all fiber optic cable maintenance loop should be installed in the MDF room in a wall-mounted enclosed standard fiber optic panel. Install a 4-inch conduit from building 150 rooms 118 to new MDF/IDF room 103.

10.0.13. In building 150 in Organization library install one (1) each 24-strand single-mode fiber optic and one (1) each 24-port single-mode fiber-optic patch panel wall mounted. Any and all fiber optic cable maintenance loop should be installed in the MDF room in a wall-mounted enclosed standard fiber optic panel.

10.0.14. Category 5E 4-pair 24 AWG copper cable plenum rated compliant types and brands shall be in accordance with MANAGEMENT CS&T Design Guidelines: 7-1-11, Revision 9.3. All Cat 5E cables will installed from rack-mounted patch panels in each floor in building 140, rooms 105B MDF, 208A IDF, 310B IDF and building 150, rooms 103 MDF/IDF, 200E IDF, 300E IDF, and 400F IDF. The maximum cable distance from a communications outlet to the termination point in a room shall not exceed 90 meters. In addition, 10 meters is allowed for interconnecting cable in the room and for station cable in the office for a total cable distance of 100 meters.

10.0.15. Plenum Rated Category 5E TIA / EIA–568–B cable with a yellow colored jacket or as requested by a customer shall be used for horizontal cabling systems. MANAGEMENT CS&T Design Guidelines: 7-1-11, Revision 9.3.

10.0.16. See attached drawings for attached horizontal plenum-rated Category 5E cable pulls required with a quantity of drops for each location proposed throughout MDF/IDF's in buildings 140 and 150.

11.0. MARKING & LABELING

11.0.1. The first outlet on the left wall will be the starting outlet. Example: Room 101D has more than one outlet the label will be 101D-1. Rotating clockwise the next outlets in room 101D will be 101D-2, 101D-3, etc. If only one outlet is in a room is labeled 101D. See MANAGEMENT CS&T Design Guidelines: 7-1-11, Revision 9.3 for complete guidelines concerning labeling.

11.0.2. Wireless Ethernet Access Point (WAP) Outlets labeling example: WAP outlet is in Room 101D. The label should read 101D-WAP. MANAGEMENT CS&T Design Guidelines: 7-1-11. Revision 9.3

11.0.3. Since WAP outlets are to be installed above drop ceilings place the appropriately labeled machine-printed label on the ceiling grid facing downward directly below the location of each WAP outlet. The label's typeface must be large enough to be read by someone standing on the floor and looking up at the label. MANAGEMENT CS&T Design Guidelines: 7-1-11. Revision 9.3

11.0.4. Labeling and marking of all infrastructure cabling will follow local MANAGEMENT Regional Campus Daytona standards. The contractor shall coordinate with the local telecommunications manager for detailed labeling and marking of all infrastructure cabling. Infrastructure cabling used for internal electronics shall follow labeling and marking standards specified in MANAGEMENT CS&T Design Guidelines: 7-1-11, Revision 9.3.

12.0. CUSTOMER PREMISE EQUIPMENT (CPE)

12.0.1. Installation of CPE shall be in accordance with the commercial installation instructions or technical instructions furnished or provided by the original equipment manufacturer for the equipment. When installation site conditions are unique the installer shall coordinate with the user to determine the configuration for installation. Instruments and equipment provided shall be MANAGEMENT approved.

12.0.2. Voice over Internet Protocol (VoIP) Phones. Installation of VOIP telephones is currently managed by the Organization (DSC) telecommunications department; MANAGEMENT CS&T shall provide assistance with configuration and assignment of ports allocations on the newly installed switching equipment. As an option, DSC may maintain a separate switching network to serve VOIP phones and may utilize the newly installed horizontal cabling to provide phone services.

13.0. FIBER OPTIC TESTS
Testing shall be performed on each segment of the cable as each splice or connector is installed to check for continuity and loss, and followed by a complete end-to-end test of the completed circuit. In addition, the following tests shall be performed on all strands of every fiber optic cable.

13.0.1. Optical Attenuation. Test using the systems optical wavelength and normal transmit level; measure with an optical power meter.

13.0.2. Optical Power Margin. Test the difference in optical levels between the normal receive level and the lowest possible receive level at which bit errors occur. The optical power margin shall be 4 dB unless the higher margin is specified by the system.

13.0.3. Cable Failure Requirements. A defective cable shall not be installed. Unless specified, failure of one or more pairs or strands shall require the Contractor to either replace the entire cable or component and retest the cable or repair those pairs or strands that fail and test pairs or strands in the cable.

13.0.4. Failure Requirements for Other Equipment. All equipment must meet the manufacturer's specification upon test completion. If any equipment fails to pass any tests, the installer/contractor shall repair/correct the failure (or replace the defective component/equipment) and re-perform all tests on that equipment. In addition, where repair or replacement invalidates or brings into doubt the results of other tests on that equipment or on associated equipment, the installer/contractor shall re-perform those tests as well.

14.0. NETWORK STAGING PLAN

14.0.1. Preface: The Organization has never had a properly designed data network and lacks a proper network backbone infrastructure. The current network electronics range in age with the newest equipment at around 2-3 years of age and the oldest equipment at nearly 15 years of age. Most of the equipment falls into the older category at 10-15 years of operation. There is no reliable network core for the campus, instead of relying on a long series of daisy-chained uplinks from a closet to closet, linking newer switches to the network through older devices that are on the verge of failure. Most of the campus network has no UPS power leaving the network susceptible to a variety of problems from power anomalies and other environmental conditions. Much of the campus lacks quality wireless network services.

14.0.2. A complete renovation of the MANAGEMENT Organization network is required. The new design will include the implementation of a true WAN router for the campus' connectivity to the main MANAGEMENT network. The new router will provide wireless management services and VPN encryption across the WAN connection back to the main campus network.

14.0.3. Two core switches with multiple single-mode interfaces rack-mounted in building 140, room 105B MDF to connect all IDF rooms. Each of the core switches will have a direct one (1-each) gigabit uplink to the WAN router. Each switch will then feed one (1-each) gigabit Ethernet connection to all IDF telecom rooms at the Daytona campus, providing redundant gigabit Ethernet links to each closet. All remaining IDF telecom distribution rooms will be equipped with new access layer switches to provide 10/100/1000/POE Ethernet service to the supported end stations. Each MDF/IDF would be equipped with a Liebert Series UPS to provide clean and reliable power. Finally, a new array of wireless network access points (WAPs) would be distributed throughout the buildings to improve signal coverage and performance for mobile users.

14.0.4. In building 140, room 105B MDF install Summit X650 24-Port 10 Gig Ethernet, Summit X480 2-Port 10 Gig 48-Port 1 Gig Ethernet, POE, Summit X460 Series 48-Port Gig Ethernet, Cisco 3900 Router, 2-Gig Ethernet, 1-Gig Ethernet WAN (Single-Mode Fiber Optics 24-Port).

14.0.5. In building 140, room 208A IDF install Summit X480 2-Port 10 Gig 48-Port 1 Gig Ethernet, POE, Summit X460 Series 48-Port Gig Ethernet (Single-Mode Fiber Optics 24-Port).

14.0.6. In building 140, room 310B IDF install Summit X480 2-Port 10 Gig 48-Port 1 Gig Ethernet, POE, Summit X460 Series 48-Port Gig Ethernet (Single-Mode Fiber Optics 24-Port).

14.0.7. In MANAGEMENT/DSC Library install Summit X480 2-Port 10 Gig 48-Ports 1 Gig Ethernet, POE, Summit X460
   Series 48-Port Gig Ethernet (Single-Mode Fiber Optics 24-Port).

14.0.8. In Bright House Metro Ethernet, building 150, room 118 install single-mode fiber optics 24-port patch panel.

14.0.9. In building 150, room 103 MDF/IDF install Summit X650 24-Port 10 Gig Ethernet, Summit X480 2-Port 10 Gig 48-Port 1 Gig Ethernet, POE, Summit X460 Series 48-Port Gig Ethernet (Single-Mode Fiber Optics 24-Port).

14.0.10. In building 150, room 200E IDF install Summit X650 24-Port 10 Gig Ethernet, Summit X480 2-Port 10 Gig 48-Port 1 Gig Ethernet, POE, Summit X460 Series 48-Port Gig Ethernet (Single-Mode Fiber Optics 24-Port).

14.0.11. In building 150, room 300E IDF install Summit X650 24-Port 10 Gig Ethernet, Summit X480 2-Port 10 Gig 48-Port 1 Gig Ethernet, POE, Summit X460 Series 48-Port Gig Ethernet (Single-Mode Fiber Optics 24-Port).

14.0.12. In building 150, room 400F IDF install Summit X650 24-Port 10 Gig Ethernet, Summit X480 2-Port 10 Gig 48-Port 1 Gig Ethernet, POE, Summit X460 Series 48-Port Gig Ethernet (Single-Mode Fiber Optics 24-Port).

14.1.0. ELECTRONICS EQUIPMENT SAFETY

14.1.2. Before installing any equipment refer to all the product documentation included with each piece of equipment. Each Engineer shall perform their duties and obligations using a reasonable degree of care, diligence, skill, and judgment. Equipment shall be handled in strict accordance with sound industry practice and its deployment and assembly shall at all times conform to the published documentation.

14.1.3. Two people are required to lift equipment chassis. Equipment must be lifted keeping a straight back, using the legs, not the back. Never attempt to lift any of the chassis by the plastic panels on the front of the chassis, or by the handles on power supplies or processor modules. These panels and handles were not designed to support the weight of the chassis.

14.1.4. Before beginning any procedures requiring access to the interior of any equipment, locate the emergency power-off switch for the room in which you are working. Do not assume that it is off, check it personally.

14.1.5. Single-mode fiber line cards are equipped with lasers that emit invisible radiation. Because invisible laser radiation may be emitted from the aperture of the port when no cable is connected, avoid exposure to laser radiation and do not stare into open apertures.

14.1.6. Confirm ESD Procedures
Provide details of any local Electrostatic Discharge procedures and precautions that must be followed at the Customer site. These could include testing personal ESD wrist straps and connecting them to their common bonding points, the wearing of additional specialist equipment such as coats and heel straps, or otherwise. Where additional equipment is required, the Customer must supply this or state it as a requirement prior to the Implementation Team's arrival so that measures may be taken to be compliant. Where no special local instructions exist, the Installation Team must use ESD wrist straps whenever handling exposed equipment that does not have a return connection to the earth; equipment removed from a sealed chassis such as interface cards and power supplies.

14.2.0. NETWORK TOPOLOGY DIAGRAMS & EQUIPMENT DETAILS
See attached diagrams clearly showing the topology of the network undergoing Implementation and detailing all components included in this document.

14.2.1. Physical Network Topology
See Attachment: Physical Network Topology.

14.2.2. Logical Network Topology
See Attachment: Logical Network Topology.

14.2.3. Equipment Floor Plan
See Attachment: Equipment Floor Plan.

14.2.4. Record Equipment Serial Numbers and Check against Delivery Documentation. Provide details of the serial numbers to be captured from field replaceable items. It is vitally important that such equipment serial numbers are tracked throughout the Implementation. The Project Manager should ensure that the serial numbers recorded in this document are compared to those recorded during Staging and supplied to the Customer Service Organization who will ensure that the records used for support purposes are updated as required and that sufficient spares are held.

14.2.5. Serial numbers are shown on equipment and invoices. The team is to record equipment serial numbers via manual inspection. Team to verify serial numbers are consistent with those shown on shipping invoices, where applicable. Any issues must be recorded on the staging hand-over certificates.

14.3.0. NETWORK TESTING

14.3.1. Complete Installation Tests
Provide a high-level list of Installation Tests to be carried out. Refer to installation tests for low-level test details. The aim of Installation Tests should primarily be to prove each piece of equipment is operational. Each test should be witnessed by the Customer and an individual test sheet must be completed.

14.3.2. Team to complete the tests indicated for each device of the respective hardware type. Installation tests for the test schedule.

14.3.3. Command output from all commands indicated in the test scripts should be recorded to the test schedules. Any failures must be indicated on the Implementation Hand-over Certificate.

Installation Tests

Test No.	Test Description

14.3.4. Complete Commissioning Tests
Provide a high-level list of Commissioning Tests to be carried out. Refer to commissioning tests for low-level test details. The aim of Commissioning Tests should primarily be to prove each site is operational. Installation Tests should not be repeated without good reason. Each test should be witnessed by the Customer and an individual test sheet must be completed.

14.3.5. Command output from all commands indicated in the test scripts should be recorded to the test schedules. Any issues must be recorded on the Implementation Hand-over Certificate.

Commissioning Tests

Test No.	Test Description

14.3.6. Command output from all commands indicated in the test scripts should be recorded to the test schedules.

14.3.7. Copy the running/startup configuration to memory and flash using the *copy running-configuration startup-configuration* and *copy running-configuration* or equivalent for different types of flash commands.

14.3.8. Add a copy of end configurations in the network ready for use test section. Any failures must be indicated on the Staging Hand-over Certificate.

Network Read for Use Tests

Test No.	Test Description

14.4.0. NETWORK DOCUMENTATION

14.4.1. Complete Implementation Record and provide any specific instructions for post-installation action.

14.4.2. Team to complete the Implementation Record for each piece of equipment and validate any tables and information in the remainder of this document not already reviewed by the actions above.

14.4.3. Team to return any spare cables, documentation, boxes, etc. leftover from installation to the Project Manager.

14.4.4. Schematic Rack Diagrams
Provide a diagram of each cabinet layout. Include the overall dimensions, positions of all supplied equipment, and dimensions of free rack space. Where DC equipment is to be used, circuit breaker diagrams should also be included here. Use WMF, or other reduced file size, format files where possible for diagrams, available via Edit - Paste Special.

14.4.5. Equipment Chassis Card Layout and Serial Numbers
Provide a diagram and/or table showing the position of each card (including card type identifier) in the equipment chassis along with serial numbers. Product IDs and Descriptions may be taken from the manufactures pricing list.

14.4.6. Device – Node Name

Device - Node name Chassis Layout and Serial Numbers

Slot	Product Description	Serial No.

14.4.7. Equipment Port Allocation and cable Schedule
Provide a site-specific cable schedule in table form detailing: from, to, interface type, cable type required, supplied by, and installed by. Include power cables particularly DC supply cabling. The link information and circuit designators may also be included in this section where applicable. Diagrams may be used in addition
to the table and should be included in this document.

14.4.8. Port Allocations. See attachment for Port and Cable Allocations.

14.4.9. Rack Designator Cable Schedule. See Attachments for Rack Designator and Cable Schedule.

14.4.10. Provide rack by rack port and cable allocations. See Attachments for Rack Designator Port and Cable Schedule.

14.4.11. Cabinet Layout and Fiber Optic Patch Cabling Infrastructure. See Attachments for Cabinet Layout and Fiber Optic Patch Cabling Infrastructure

14.4.12. Labeling Formats
See attachments for details of cabinets, cables, etc. should be labeled.

14.4.13. Equipment Labeling

14.4.14. Sticker labels showing the chassis name should be applied to the front and rear of the equipment, Example: PE01.

14.4.15. Power Supply Labeling.

14.4.16. None where there is only a single supply.

14.4.17. Where dual power supplies exist no additional labels to chassis markings need to be made as indicators are provided as part of the chassis. Additional labels may be applied to re-enforce these.

14.4.18. AC/DC Breakers should have magnetic labels installed beneath the breakers indicating the device name and power supply indicator. Example: CE02 A.

14.4.19. Cabinet Labeling.

14.4.20. Labels showing the following should be applied at the top of the front/rear of the cabinets.

14.4.21. The front door of the rack must display a label "Front" and Laser warning labels. The rear must display "Back" with Laser warning labels.

XXXX Rack Front/Back Labeling Format

Site Name	Manufacture	Rack
(Site Name)	(XXXX Name)	X

14.4.21. Cable Labeling.

14.4.22. Cable labels should be applied at both ends of each cable, within 0.5m of the termination point. They should be permanently fixed around (not to) the cable in printed ink (6 point minimum).

14.4.23. For fiber patch cables, Kroy (wrap around) labels should be used with text in the following form: (source rack no) (source device or patch panel name) (slot/port/interface) to (destination can rack no) (destination device or patch panel name) (slot/port/interface).

14.4.24. For UTP patch cables to/from the 6509 equipment that terminates at the patch panel labels should be of the form: (device name) S (slot no.) / (0port no).

14.4.25. Equipment Software Requirements
Provide details of the revisions of software, firmware, boot code, and microcode to be installed during Implementation.

Equipment Software IOS Requirements

Equipment	Software/Hardware/Firmware/Boot code Revisions	Image Name	Release Status

14.4.26. Equipment Basic Setup Procedures
Provide details of any equipment specifics that are required immediately after power-up to ensure no configuration anomalies are present before the configuration is applied if such instructions are not in manuals shipped with documentation. Likewise for any software upgrade procedures that may be required. Note that if Network Staging has been completed such procedures may not be necessary and may be omitted.

14.4.27. Configuration Clearing
Provide instructions for defaulting or clearing configuration settings.

14.4.28. Before configuring and adding the node to an existing network, it is recommended to clear the existing configuration of the node. Use the following procedure to clear the configuration of an XXXXXX:

Step 1.
Step 2.
Step 3.

14.4.29. Software Upgrade
Provide instructions to upgrade or downgrade software, firmware, boot code, or otherwise.

14.4.30. Equipment Configurations
Provide details of configurations to be loaded to each device. State clear instructions as to how the configuration is to be loaded, i.e., write run start, no shutdown, etc. commands should be documented. Likewise for any caveats that should be observed while inputting configurations. NMS information including disk partitioning, file setup, etc. should also be included here as should point code and other device configuration information.

14.4.31. Network Addressing Schemes
Provide point code details for signaling and related equipment, IP addressing ranges, naming conventions, node numbering, etc.

14.4.32. Naming Conventions
Provide details of the naming conventions used in the network/this site.

14.4.33. Devices are named using the following convention:

14.4.34. IP Addressing Scheme
Provide details of the IP addressing scheme used at this site. Refer to Network/ IP Addressing Plan.

IP Addressing Scheme

IP Address	Device	Device Type	Port	Remarks

14.4.35. Point Code Addressing Scheme
Provide details of the point code addressing scheme used at this site.

Point Code Addressing Schemes

Exchange	Network Indicator	Point Code (Decimal)	Point Code (3-8-3)

14.4.36. Device – Node name Configuration
Provide go-to-site device configuration details and the order in which commands should be executed or add as objects text files that contain the configurations.

14.4.37. Baseline:
NMS System - Hostname Configuration
Provide details of the NMS system configuration for each host in text form or added as a text file object.

14.4.38. End of Implementation Configurations
Provide the configurations present on the devices at the end of implementation, if different from those above. List in-line text form or add as objects/text files that contain the configurations.

14.4.39. Installation Tests
Provide a composite set of Tests made up of individual Tests taken from Test library. After new Tests have been written they should be published individually on the PE document server. The aim of Installation

Tests should primarily be to prove each piece of equipment is operational. Tests should not be given Table numbers. Each test should start on a new page. Test output should be added as in-line text or objects/text files that contain the output.

14.4.40. Commissioning Test
Provide a composite set of Tests made up of individual Tests taken from Test library. After new Tests have been written they should be published individually on the PE document server. The aim of Commissioning Tests should primarily be to prove each site is operational. Installation Tests should not be repeated without good reason. Tests should not be given Table numbers. Each test should start on a new page. Test output should be added as in-line text or objects/text files that contain the output.

14.4.41. Network Ready for Use Test
Provide a composite set of Tests made up of individual Tests taken from Test Library. After new Tests have been written they should be published individually on the PE document server. The aim of Ready for Use Tests should primarily be to prove the network as a single entity is operational and manageable and which should be possible using Standard Tests. Non-Standard Tests may be discussed with the Customer, as necessary, and costs in the SOW. Non-Standard Tests are generally Tests that could have a subjective result, e.g., network round-trip-delay, buffer performance, throughput, call setup time, network/node limits and characteristics when under stress/loading, or of the like. Installation and Commissioning Tests should not be repeated without good reason. Tests should not be given Table numbers. Each test should start on a new page. Test output should be added as in-line text or objects/text files that contain the output.

15.0 PHOTOGRAPHS
Provide, where possible, photographic evidence of the Implementation. Photographs should be inserted in JPEG, or similar (not bitmap), format. Include cabinet front and rear views as a minimum. Note that many customers do not permit photographs to be taken in their communications rooms without prior permission. Verification should be obtained by the Implementation Engineer and Project Manager prior to site attendance, where necessary this should also be further verified by the on-site engineer).

16.0 PROJECT DRAWINGS

16.0.1. Upon project completion, the team leader annotates two sets of project drawings.
The team leader leaves one set with the telecommunications manager and sends the other set to the communications planning office. Drawings are annotated copies of project drawings compiled by the team leader on completion of the installation of a project at a particular location.

16.0.2. They are commonly called "as installed" drawings and annotated to reflect the as-installed conditions that vary from the actual project drawings furnished by the team leader to the telecommunications and planning office.

Telecommunications Staging Hand-over

3.22 Telecommunications Staging Hand-over

Certificate type: (Description of Sign-off)

System description: (Customer Name/Project Name)

Document version: Version (x.y)

Network staging for the (Customer Project) network has been completed to the satisfaction of (Customer). All Installation, Commissioning and, where appropriate, Network Ready for Use Tests have been completed and recorded.

Please note any comments below.

For and on behalf of MANAGEMENT Telecommunications and Networking Group:

Name: **Signature:** **Date:**
(Print) (Sign)

For and on behalf of (Customer):

Name: **Signature:** **Date:**
(Print) (Sign)

Comments, variations or caveats:

Note: This ACCEPTANCE CERTIFICATE is to be completed and signed and returned to the Project Manager.

ATTACHMENTS

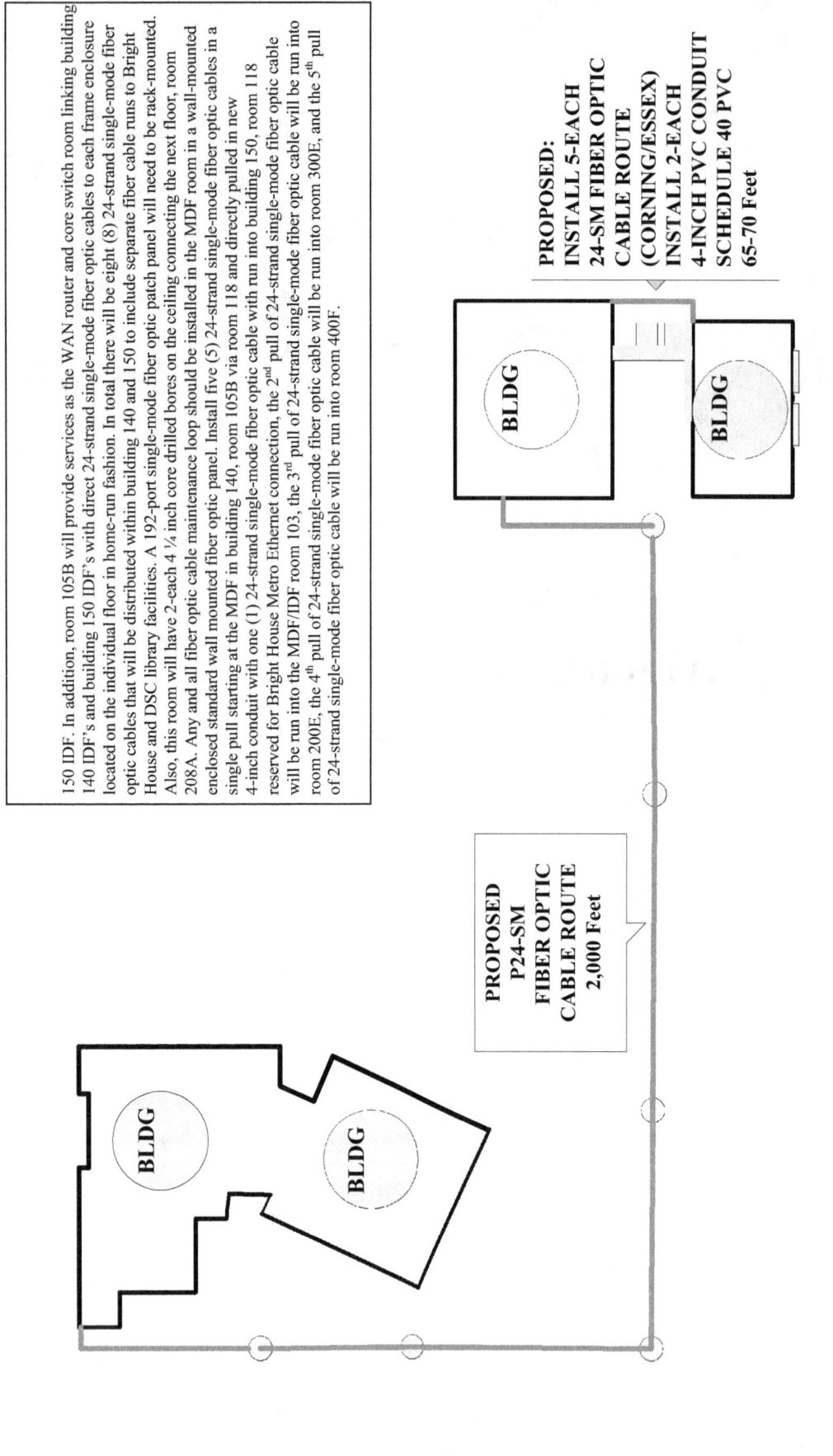

150 IDF. In addition, room 105B will provide services as the WAN router and core switch room linking building 140 IDF's and building 150 IDF's with direct 24-strand single-mode fiber optic cables to each frame enclosure located on the individual floor in home-run fashion. In total there will be eight (8) 24-strand single-mode fiber optic cables that will be distributed within building 140 and 150 to include separate fiber cable runs to Bright House and DSC library facilities. A 192-port single-mode fiber optic patch panel will need to be rack-mounted. Also, this room will have 2-each 4 ¼ inch core drilled bores on the ceiling connecting the next floor, room 208A. Any and all fiber optic cable maintenance loop should be installed in the MDF room in a wall-mounted enclosed standard wall mounted fiber optic panel. Install five (5) 24-strand single-mode fiber optic cables in a single pull starting at the MDF in building 140, room 105B via room 118 and directly pulled in new 4-inch conduit with one (1) 24-strand single-mode fiber optic cable with run into building 150, room 118 reserved for Bright House Metro Ethernet connection, the 2nd pull of 24-strand single-mode fiber optic cable will be run into the MDF/IDF room 103, the 3rd pull of 24-strand single-mode fiber optic cable will be run into room 200E, the 4th pull of 24-strand single-mode fiber optic cable will be run into room 300E, and the 5th pull of 24-strand single-mode fiber optic cable will be run into room 400F.

**PROPOSED:
INSTALL 5-EACH
24-SM FIBER OPTIC
CABLE ROUTE
(CORNING/ESSEX)
INSTALL 2-EACH
4-INCH PVC CONDUIT
SCHEDULE 40 PVC
65-70 Feet**

**PROPOSED
P24-SM
FIBER OPTIC
CABLE ROUTE
2,000 Feet**

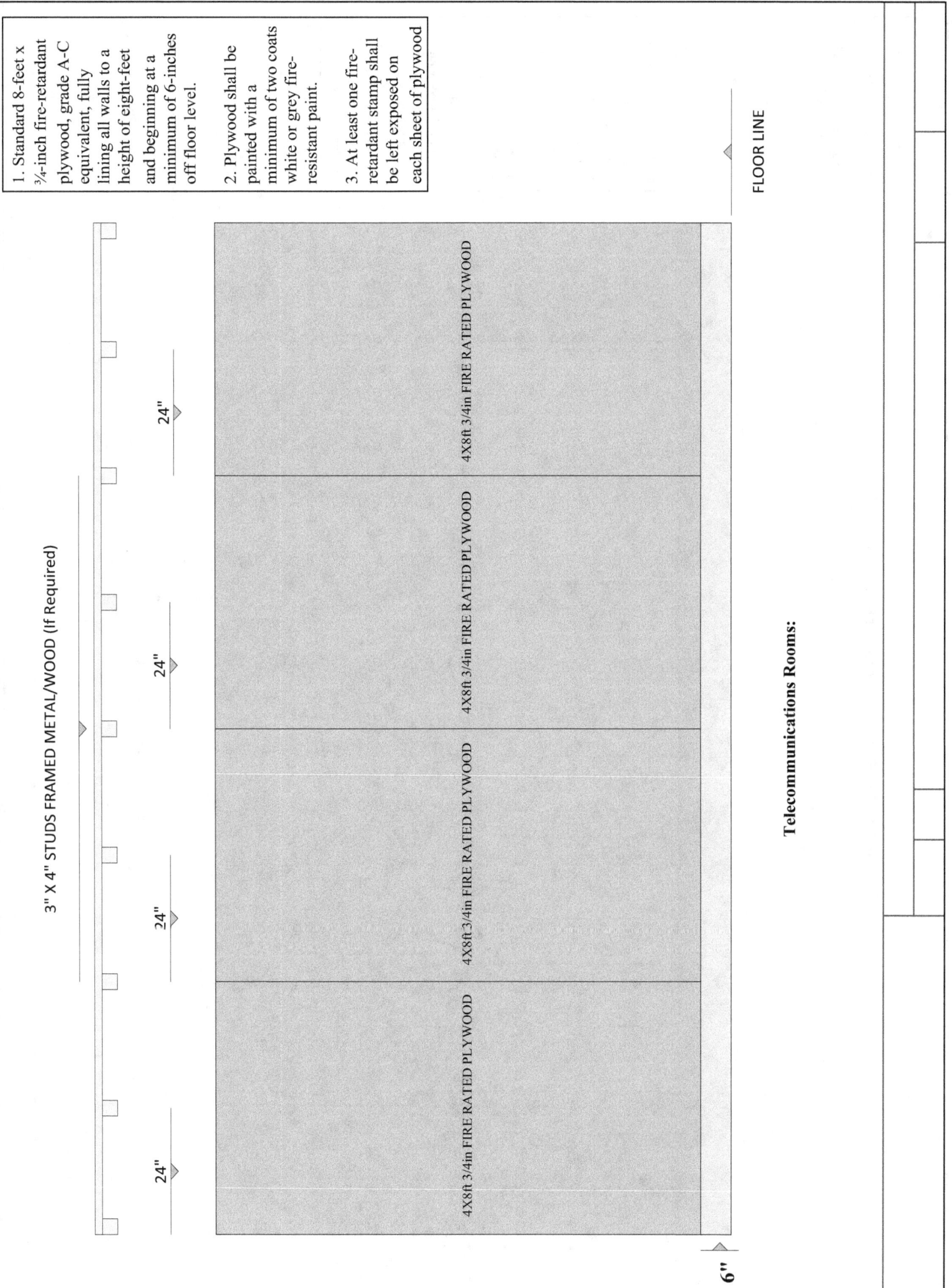

Telecommunications Rooms:

1. Standard 8-feet x ¾-inch fire-retardant plywood, grade A-C equivalent, fully lining all walls to a height of eight-feet and beginning at a minimum of 6-inches off floor level.

2. Plywood shall be painted with a minimum of two coats white or grey fire-resistant paint.

3. At least one fire-retardant stamp shall be left exposed on each sheet of plywood.

4. In MDF/IDF telecom rooms with windows build 3" X 4" frame metal/wood stud frame.

5. Install ¾-inch fire rated plywood to exactly fit the entire framed window area in one piece if possible to allow for easy access.

6. Paint back of ¾-inch fire rated plywood with white paint to match exterior décor.

7. Do not paint windows leave as-is to match facilities aesthetics.

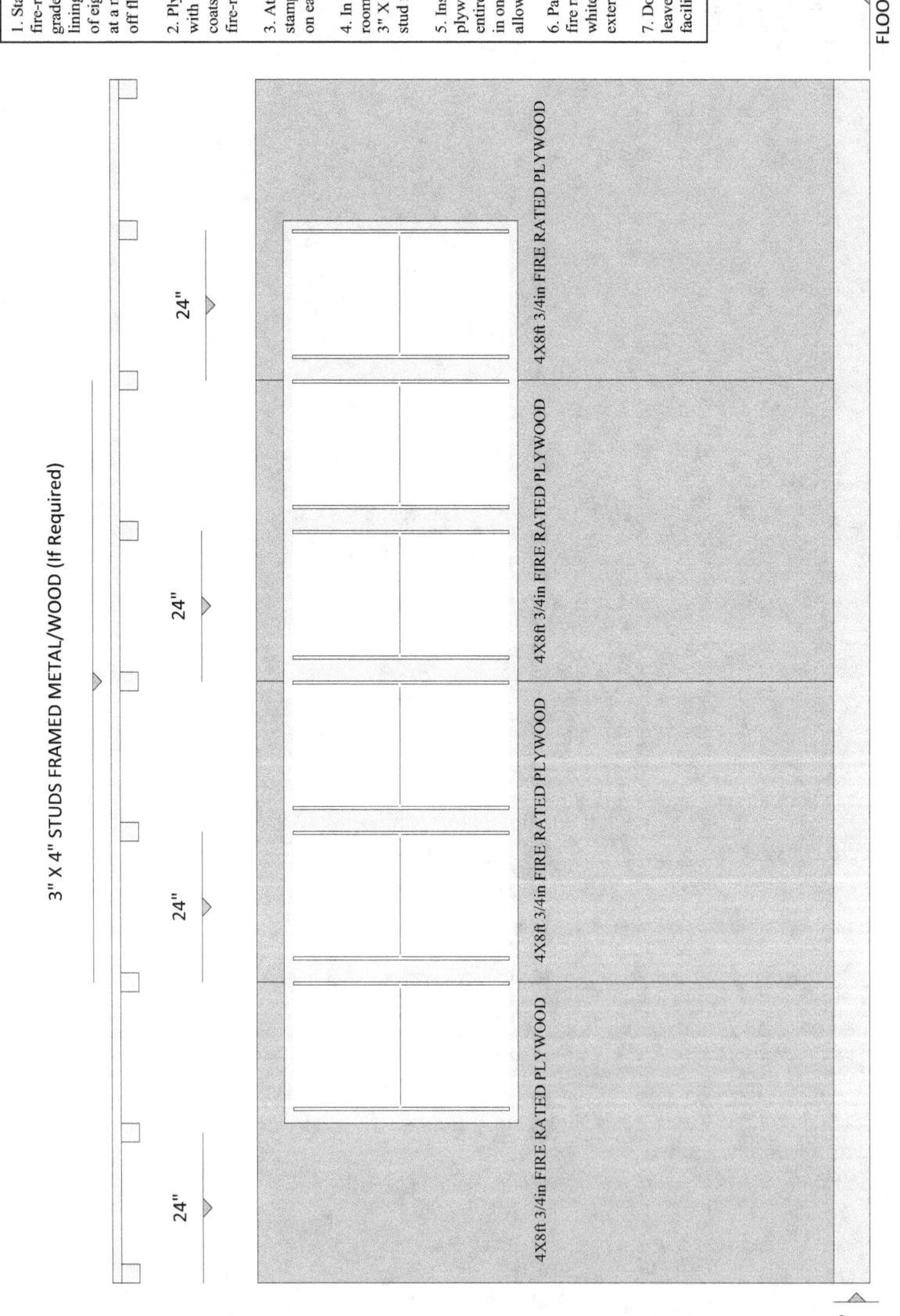

3" X 4" STUDS FRAMED METAL/WOOD (If Required)

24"
24"
24"
24"

4X8ft 3/4in FIRE RATED PLYWOOD
4X8ft 3/4in FIRE RATED PLYWOOD
4X8ft 3/4in FIRE RATED PLYWOOD
4X8ft 3/4in FIRE RATED PLYWOOD

FLOOR LINE

6"

Telecommunications Rooms:

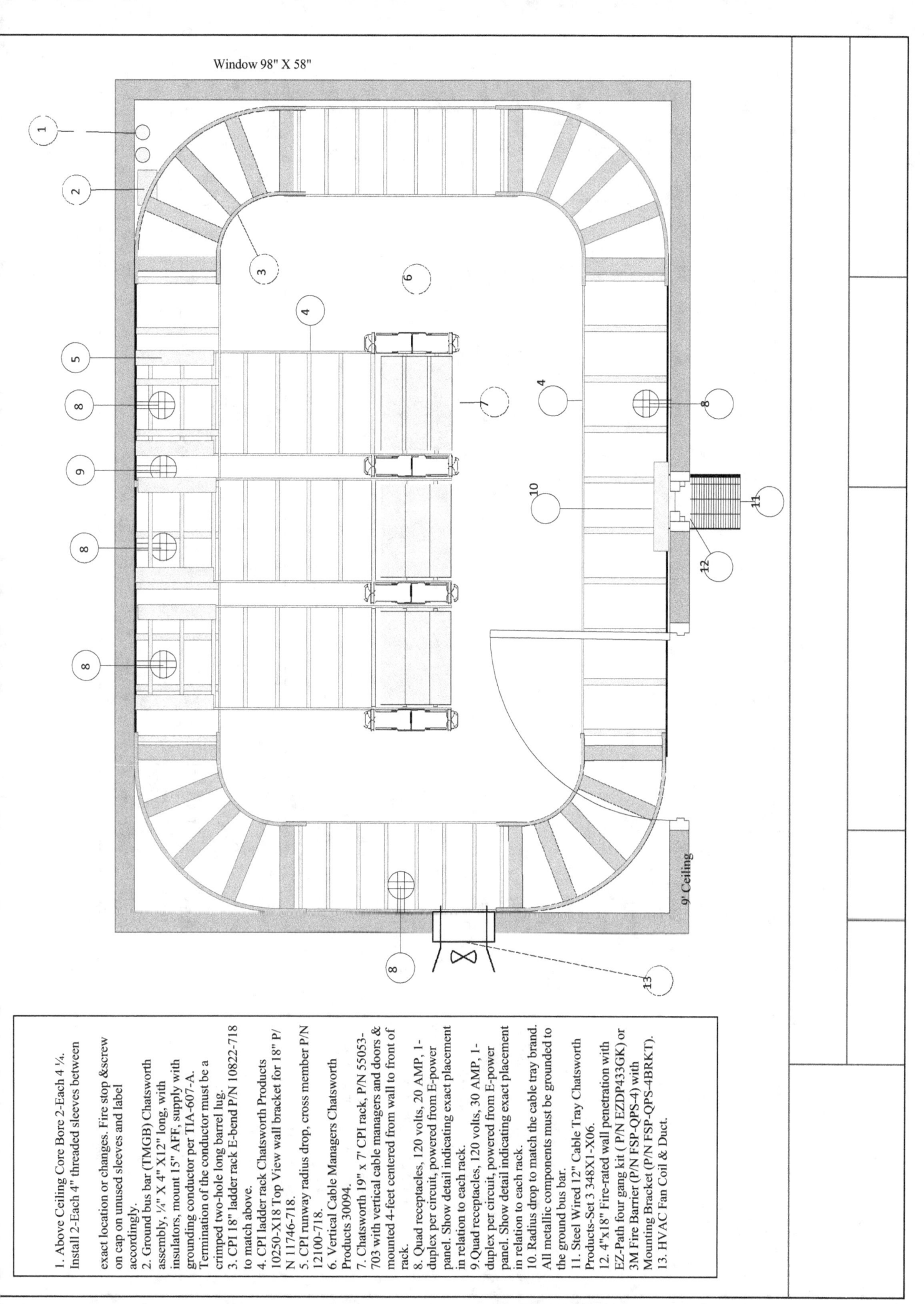

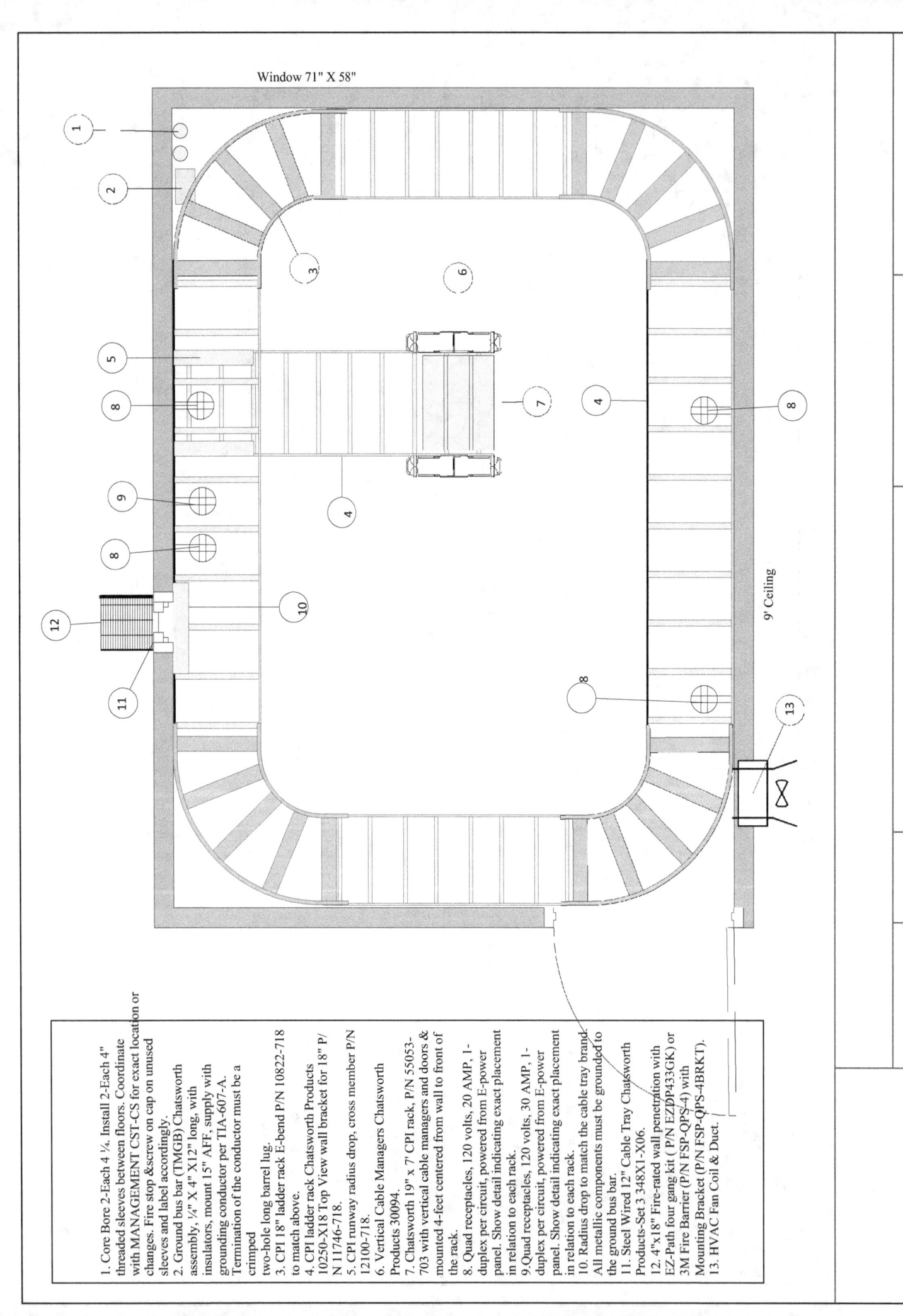

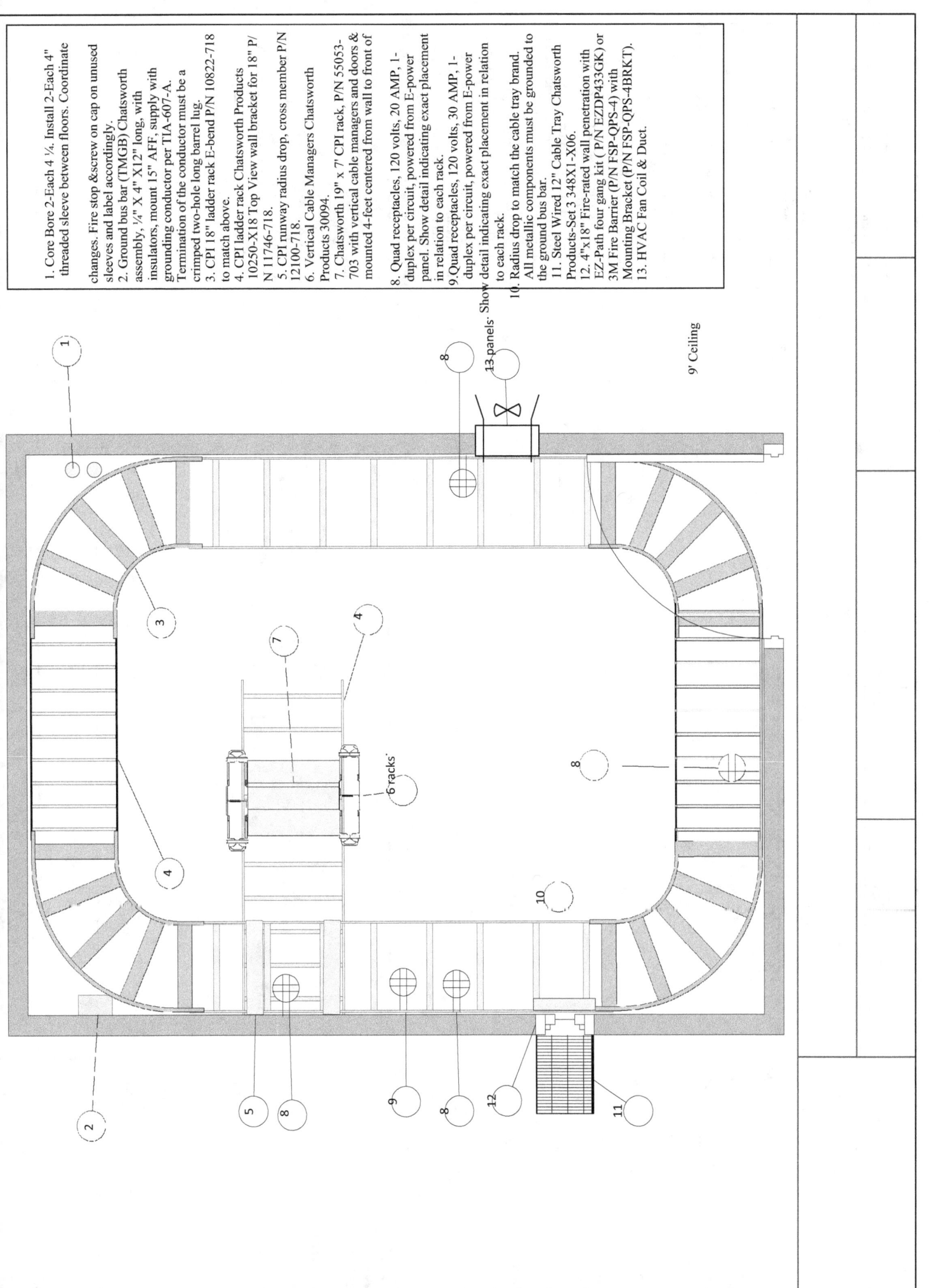

1. Core Bore 2-Each 4 ¼. Install 2-Each 4" threaded sleeve between floors. Coordinate changes. Fire stop &screw on cap on unused sleeves and label accordingly.
2. Ground bus bar (TMGB) Chatsworth assembly, ¼" X 4" X12" long, with insulators, mount 15" AFF, supply with grounding conductor per TIA-607-A. Termination of the conductor must be a crimped two-hole long barrel lug.
3. CPI 18" ladder rack E-bend P/N 10822-718 to match above.
4. CPI ladder rack Chatsworth Products 10250-X18 Top View wall bracket for 18" P/N 11746-718.
5. CPI runway radius drop, cross member P/N 12100-718.
6. Vertical Cable Managers Chatsworth Products 30094.
7. Chatsworth 19" x 7" CPI rack, P/N 55053-703 with vertical cable managers and doors & mounted 4-feet centered from wall to front of panels. Show detail indicating exact placement in relation to each rack.
8. Quad receptacles, 120 volts, 20 AMP, 1-duplex per circuit, powered from E-power panel. Show detail indicating exact placement in relation to each rack.
9. Quad receptacles, 120 volts, 30 AMP, 1-duplex per circuit, powered from E-power panel. Show detail indicating exact placement in relation to each rack.
10. Radius drop to match the cable tray brand. All metallic components must be grounded to the ground bus bar.
11. Steel Wired 12" Cable Tray Chatsworth Products-Set 3 348X1-X06.
12. 4"x18" Fire-rated wall penetration with EZ-Path four gang kit (P/N EZDP433GK) or 3M Fire Barrier (P/N FSP-QPS-4) with Mounting Bracket (P/N FSP-QPS-4BRKT).
13. HVAC Fan Coil & Duct.

9' Ceiling

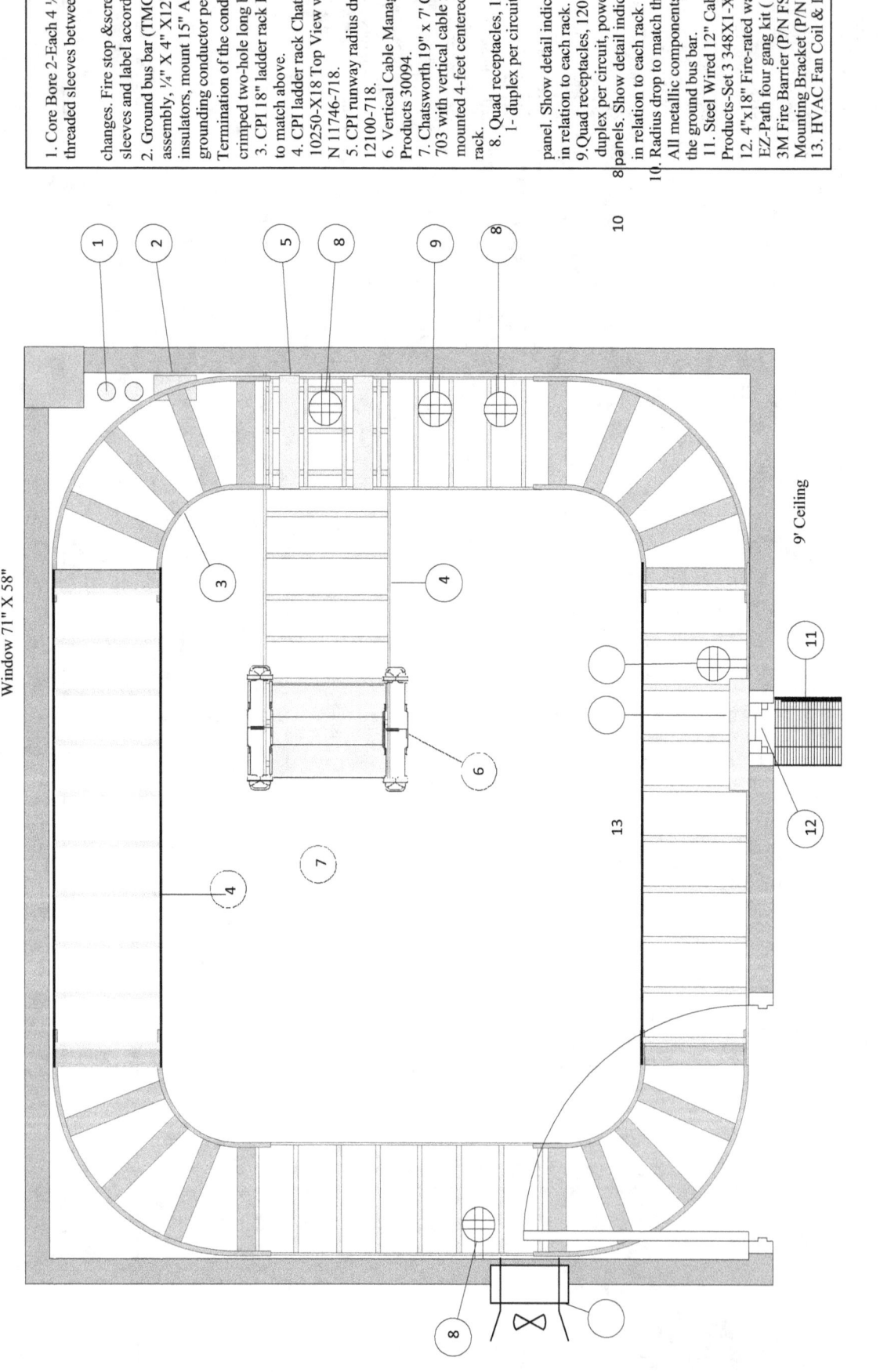

1. Core Bore 2-Each 4 ¼. Install 2-Each 4" threaded sleeves between floors. Coordinate changes. Fire stop &screw on cap on unused sleeves and label accordingly.
2. Ground bus bar (TMGB) Chatsworth assembly, ¼" X 4" X12" long, with insulators, mount 15" AFF, supply with grounding conductor per TIA-607-A. Termination of the conductor must be a crimped two-hole long barrel lug.
3. CPI 18" ladder rack E-bend P/N 10822-718 to match above.
4. CPI ladder rack Chatsworth Products 10250-X18 Top View wall bracket for 18" P/N 11746-718.
5. CPI runway radius drop, cross member P/N 12100-718.
6. Vertical Cable Managers Chatsworth Products 30094.
7. Chatsworth 19" x 7" CPI rack, P/N 55053-703 with vertical cable managers and doors & mounted 4-feet centered from wall to front of rack.
8. Quad receptacles, 120 volts, 20 AMP, 1- duplex per circuit, powered from E-power panel. Show detail indicating exact placement in relation to each rack.
9. Quad receptacles, 120 volts, 30 AMP, 1-duplex per circuit, powered from E-power panels. Show detail indicating exact placement in relation to each rack.
10. Radius drop to match the cable tray brand. All metallic components must be grounded to the ground bus bar.
11. Steel Wired 12" Cable Tray Chatsworth Products-Set 3 348X1-X06.
12. 4"x18" Fire-rated wall penetration with EZ-Path four gang kit (P/N EZDP433GK) or 3M Fire Barrier (P/N FSP-QPS-4) with Mounting Bracket (P/N FSP-QPS-4BRKT).
13. HVAC Fan Coil & Duct.

Window 71" X 58"

9' Ceiling

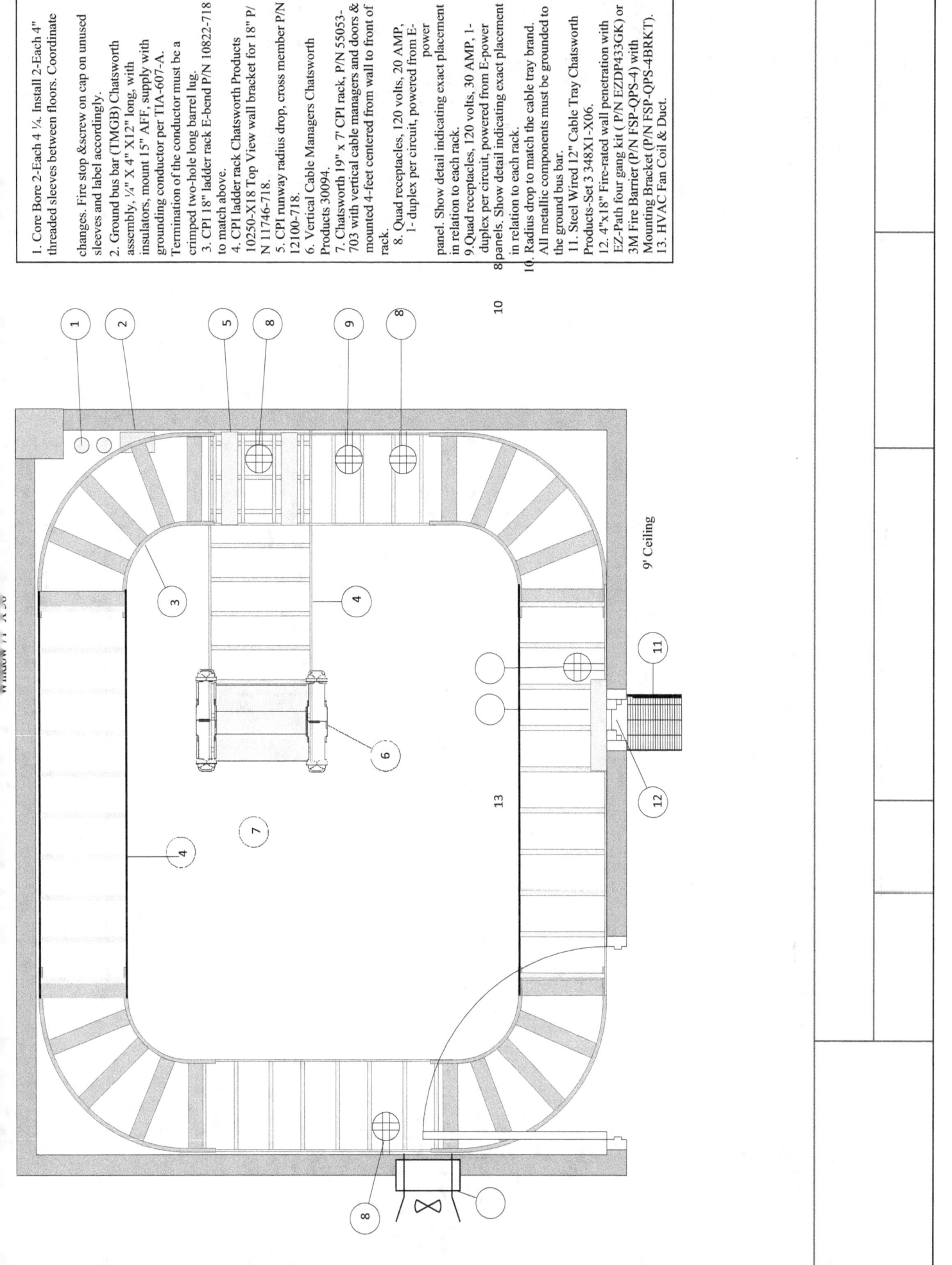

1. Core Bore 2-Each 4 ¼. Install 2-Each 4" threaded sleeves between floors. Coordinate changes. Fire stop &screw on cap on unused sleeves and label accordingly.
2. Ground bus bar (TMGB) Chatsworth assembly, ¼" X 4" X12" long, with insulators, mount 15" AFF, supply with grounding conductor per TIA-607-A. Termination of the conductor must be a crimped two-hole long barrel lug.
3. CPI 18" ladder rack E-bend P/N 10822-718 to match above.
4. CPI ladder rack Chatsworth Products 10250-X18 Top View wall bracket for 18" P/N 11746-718.
5. CPI runway radius drop, cross member P/N 12100-718.
6. Vertical Cable Managers Chatsworth Products 30094.
7. Chatsworth 19" x 7" CPI rack, P/N 55053-703 with vertical cable managers and doors & mounted 4-feet centered from wall to front of rack.
8. Quad receptacles, 120 volts, 20 AMP, 1- duplex per circuit, powered from E-power panel. Show detail indicating exact placement in relation to each rack.
9. Quad receptacles, 120 volts, 30 AMP, 1- duplex per circuit, powered from E-power panels. Show detail indicating exact placement in relation to each rack.
10. Radius drop to match the cable tray brand. All metallic components must be grounded to the ground bus bar.
11. Steel Wired 12" Cable Tray Chatsworth Products-Set 3 348X1-X06.
12. 4"x18" Fire-rated wall penetration with EZ-Path four gang kit (P/N EZDP433GK) or 3M Fire Barrier (P/N FSP-QPS-4) with Mounting Bracket (P/N FSP-QPS-4BRKT).
13. HVAC Fan Coil & Duct.

9' Ceilng

1. Core Bore 2-Each 4 ¼. Install 2-Each 4" threaded sleeves between floors. Coordinate changes. Fire stop &screw on cap on unused sleeves and label accordingly.
2. Ground bus bar (TMGB) Chatsworth assembly, ¼" X 4" X12" long, with insulators, mount 15" AFF, supply with grounding conductor per TIA-607-A. Termination of the conductor must be a crimped two-hole long barrel lug.
3. CPI 18" ladder rack E-bend P/N 10822-718 to match above.
4. CPI ladder rack Chatsworth Products 10250-X18 Top View wall bracket for 18" P/N 11746-718.
5. CPI runway radius drop, cross member P/N 12100-718.
6. Vertical Cable Managers Chatsworth Products 30094.
7. Chatsworth 19" x 7 CPI rack, P/N 55053-703 with vertical cable managers and doors & mounted 4-feet centered from wall to front of rack.
8. Quad receptacles, 120 volts, 20 AMP, 1- duplex per circuit, powered from E-power panel. Show detail indicating exact placement in relation to each rack.
9. Quad receptacles, 120 volts, 30 AMP, 1-duplex per circuit, powered from E-power panels. Show detail indicating exact placement in relation to each rack.
10. Radius drop to match the cable tray brand. All metallic components must be grounded to the ground bus bar.
11. Steel Wired 12" Cable Tray Chatsworth Products-Set 3 348X1-X06.
12. 4"x18" Fire-rated wall penetration with EZ-Path four gang kit (P/N EZDP433GK) or 3M Fire Barrier (P/N FSP-QPS-4) with Mounting Bracket (P/N FSP-QPS-4BRKT).
13. HVAC Fan Coil & Duct.

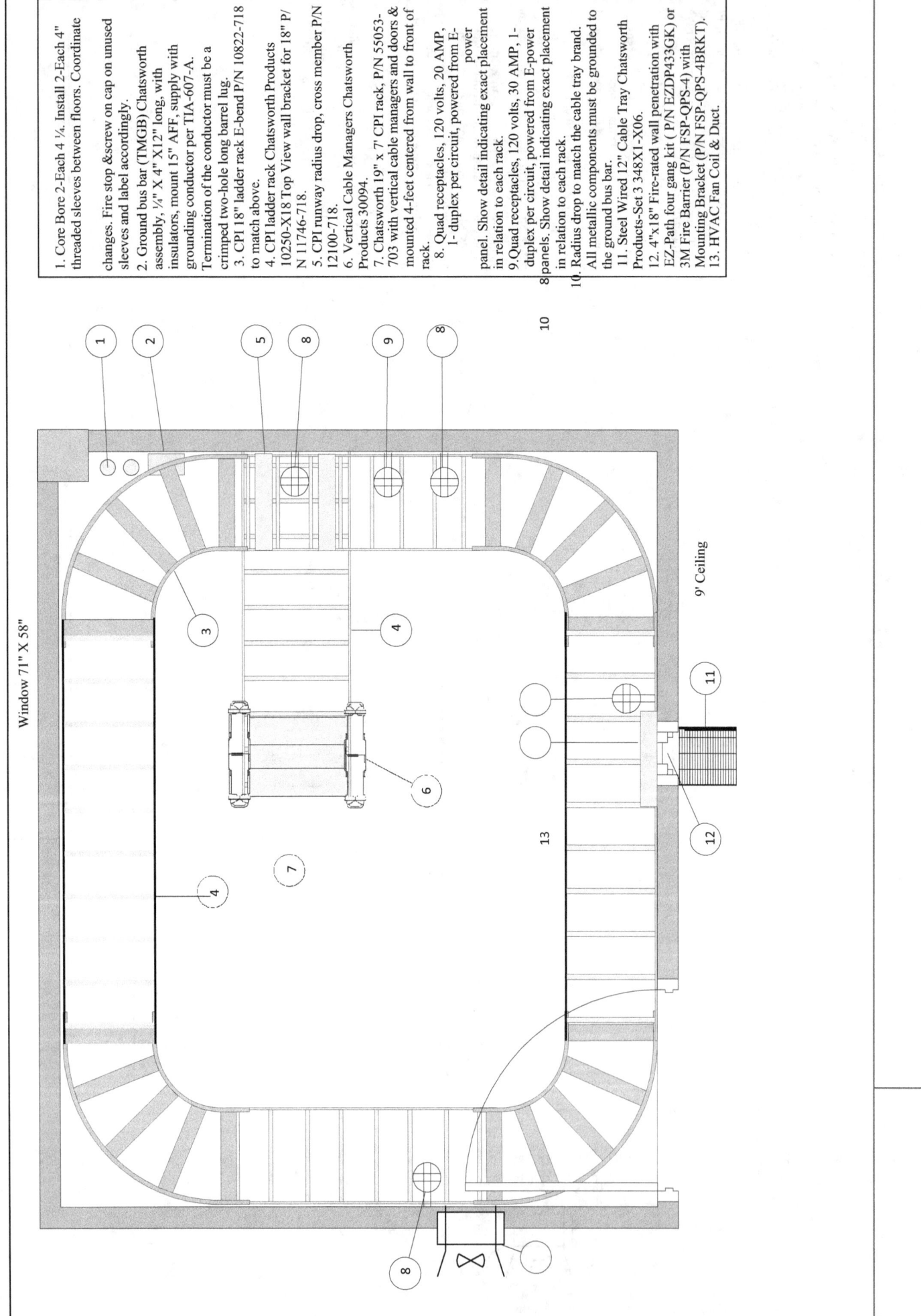

Window 71" X 58"

9' Ceiling

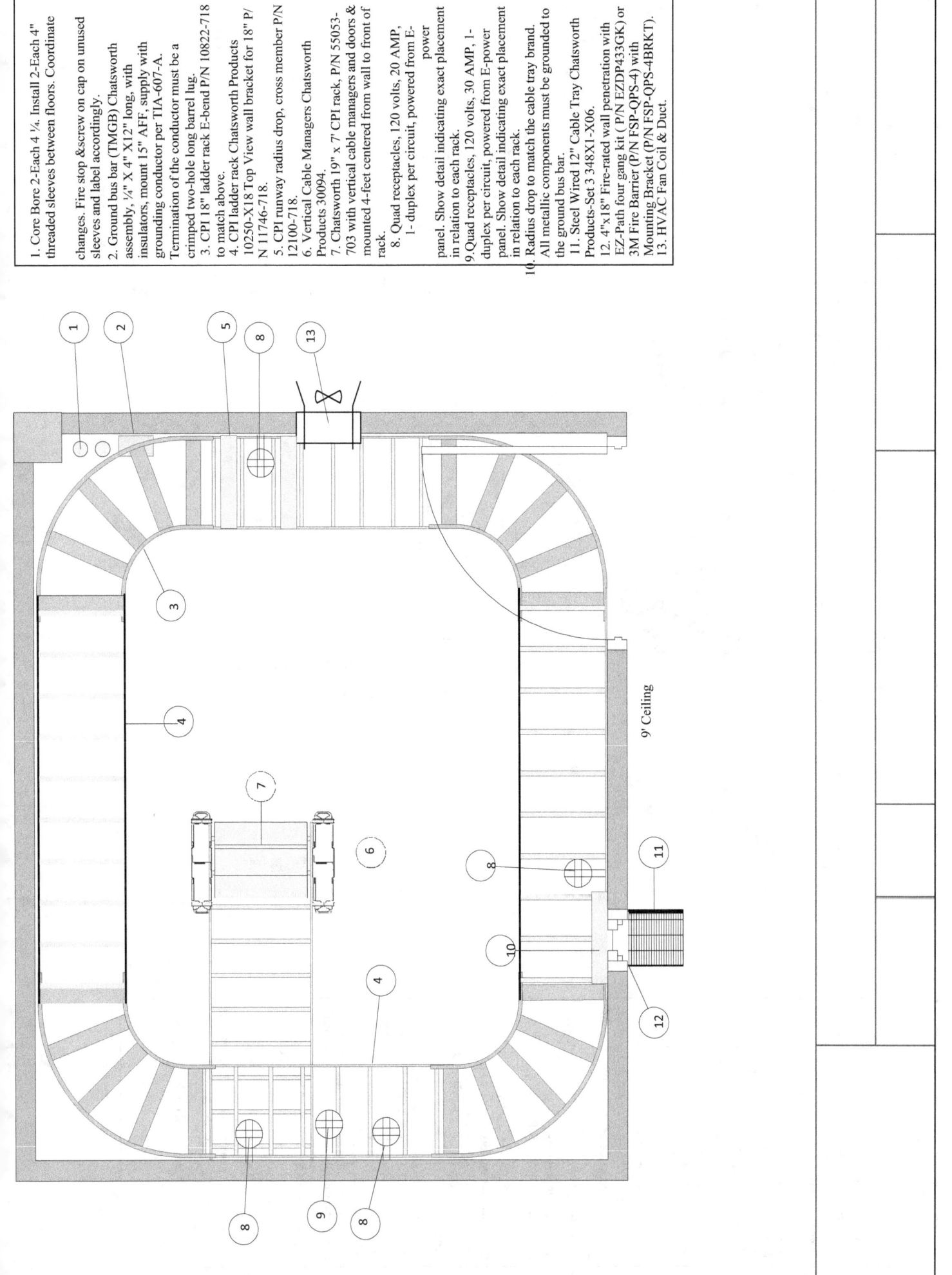

1. Core Bore 2-Each 4 ¼. Install 2-Each 4" threaded sleeves between floors. Coordinate changes. Fire stop &screw on cap on unused sleeves and label accordingly.
2. Ground bus bar (TMGB) Chatsworth assembly, ¼" X 4" X12" long, with insulators, mount 15" AFF, supply with grounding conductor per TIA-607-A. Termination of the conductor must be a crimped two-hole long barrel lug.
3. CPI 18" ladder rack E-bend P/N 10822-718 to match above.
4. CPI ladder rack Chatsworth Products 10250-X18 Top View wall bracket for 18" P/N 11746-718.
5. CPI runway radius drop, cross member P/N 12100-718.
6. Vertical Cable Managers Chatsworth Products 30094.
7. Chatsworth 19" x 7" CPI rack, P/N 55053-703 with vertical cable managers and doors & mounted 4-feet centered from wall to front of rack.
8. Quad receptacles, 120 volts, 20 AMP, 1- duplex per circuit, powered from E-power panel. Show detail indicating exact placement in relation to each rack.
9. Quad receptacles, 120 volts, 30 AMP, 1- duplex per circuit, powered from E-power panel. Show detail indicating exact placement in relation to each rack.
10. Radius drop to match the cable tray brand. All metallic components must be grounded to the ground bus bar.
11. Steel Wired 12" Cable Tray Chatsworth Products-Set 3 348X1-X06.
12. 4"x18" Fire-rated wall penetration with EZ-Path four gang kit (P/N EZDP433GK) or 3M Fire Barrier (P/N FSP-QPS-4) with Mounting Bracket (P/N FSP-QPS-4BRKT).
13. HVAC Fan Coil & Duct.

9' Ceiling

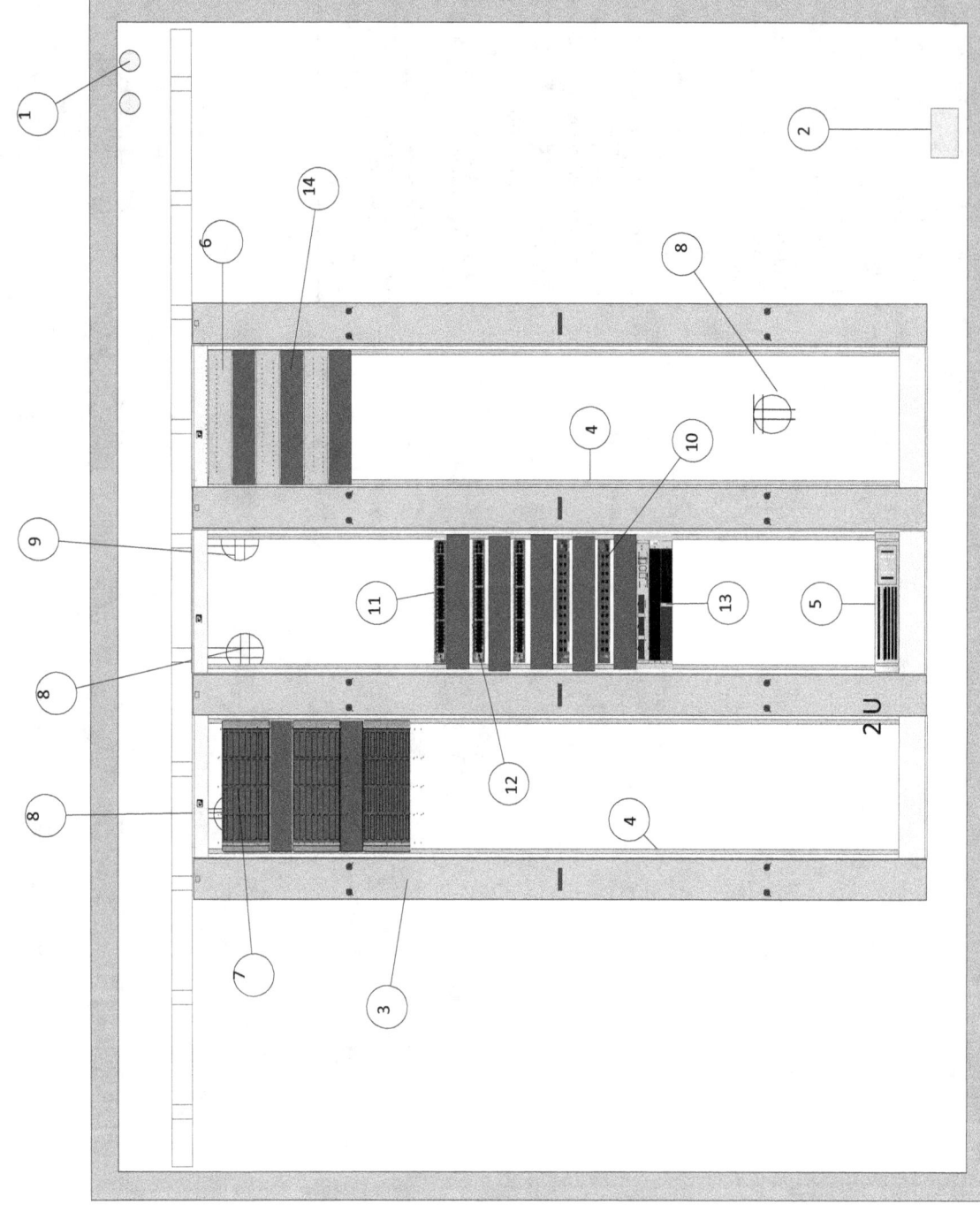

Keyed Notes: Accomplished by employees or General Contractor.
1. Above Ceiling Core Bore 2-Each 4 ¼". Install 2-Each 4" threaded sleeves between exact location or changes. Fire stop &screw on cap on unused sleeves and label accordingly.
2. Ground bus bar (TMGB) Chatsworth assembly, ¼" X 4" X12" long, with insulators, mount 15" AFF, supply with grounding conductor per TIA-607-A. Termination of conductor must be a crimped two-hole long barrel lug.
3. Vertical Cable Managers Chatsworth Products 30092-508.
4. Chatsworth 19" x 7' CPI rack, P/N 55053-703 with vertical cable managers and doors & mounted 4-feet centered from wall to front of rack.
5. Liebert GXT3 On-Line UPS, 500-3000VA
6. Hubble UDX48E 48 PORT XCELERATOR JACK PANEL
7. Corning EDGE-04U-FP Pretium EDGE® Solutions Housing, Fixed, 4 rack units, hold 32 Pretium EDGE Solutions Modules or Panels (with field removable strain-relief plate).
8. Quad receptacles, 120 volts, 20 AMP, 1-duplex per circuit, powered from E-power panel. Show detail indicating exact placement in relation to each rack.
9. Quad receptacles, 120 volts, 30 AMP, 1-duplex per circuit, powered from E-power panel. Show detail indicating exact placement in relation to each rack.
10. Summit X650 24-Port 10 Gig Ethernet,
11. Summit X480 2-Port 10 Gig 48-Port 1 Gig Ethernet, POE,
12. Summit X460 Series 48-Port Gig Ethernet.
13. Cisco 3900 Router, 2-Gig Ethernet, 1-Gig Ethernet WAN (Single-Mode Fiber Optics 24-Port)
14. Chatsworth Horizontal Cable Managers Chatsworth13930

9' Original Drop Ceiling

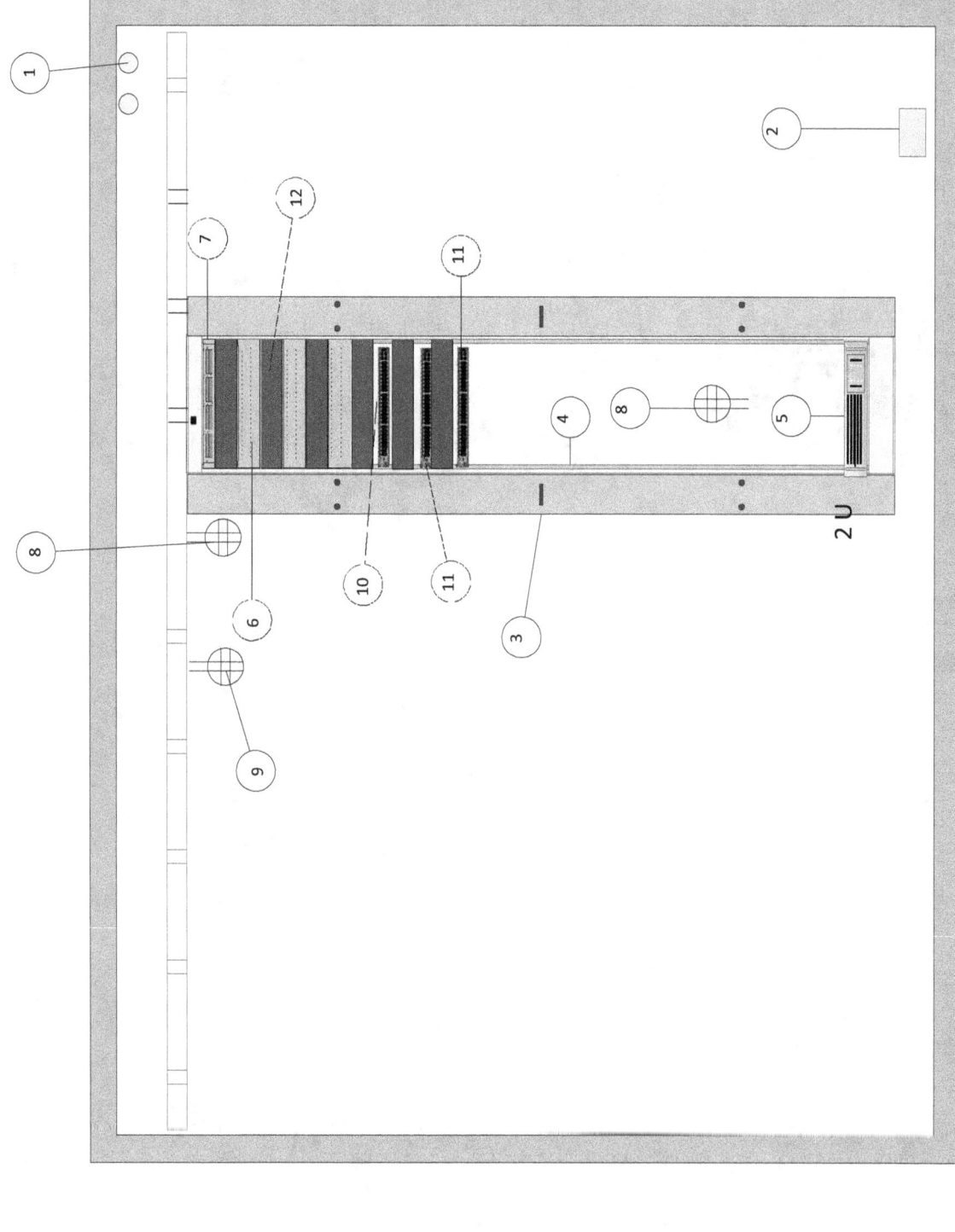

9 Original Drop Ceiling

1. Above Ceiling Core Bore 2-Each 4 ¼. Install 2-Each 4" threaded sleeves between exact location or changes. Fire stop &screw on cap on unused sleeves and label accordingly.
2. Ground bus bar (TMGB) Chatsworth assembly, ¼" X 4" X12" long, with insulators, mount 15" AFF, supply with grounding conductor per TIA-607-A. Termination of conductor must be a crimped two-hole long barrel lug.
3. Vertical Cable Managers Chatsworth Products 30092-508.
4. Chatsworth 19" x 7" CPI rack, P/N 55053-703 with vertical cable managers and doors & mounted 4-feet centered from wall to front of rack.
5. Liebert GXT3 On-Line UPS, 500-3000VA
6. Hubble UDX48E 48 PORT XCELERATOR JACK PANEL
7. Corning EDGE-01U-FP Pretium EDGE® Solutions Housing, Fixed, 1 rack units, hold 8 Pretium EDGE Solutions Modules or Panels (with field removable strain-relief plate).
8. Quad receptacles, 120 volts, 20 AMP, 1-duplex per circuit, powered from E-power panel. Show detail indicating exact placement in relation to each rack.
9.Quad receptacles, 120 volts, 30 AMP, 1-duplex per circuit, powered from E-power panel. Show detail indicating exact placement in relation to each rack.
10. Summit X480 2-Port 10 Gig 48-Port 1 Gig Ethernet, POE,
11. Summit X460 Series 48-Port Gig Ethernet.
12. Chatsworth Horizontal Cable Managers Chatsworth13930

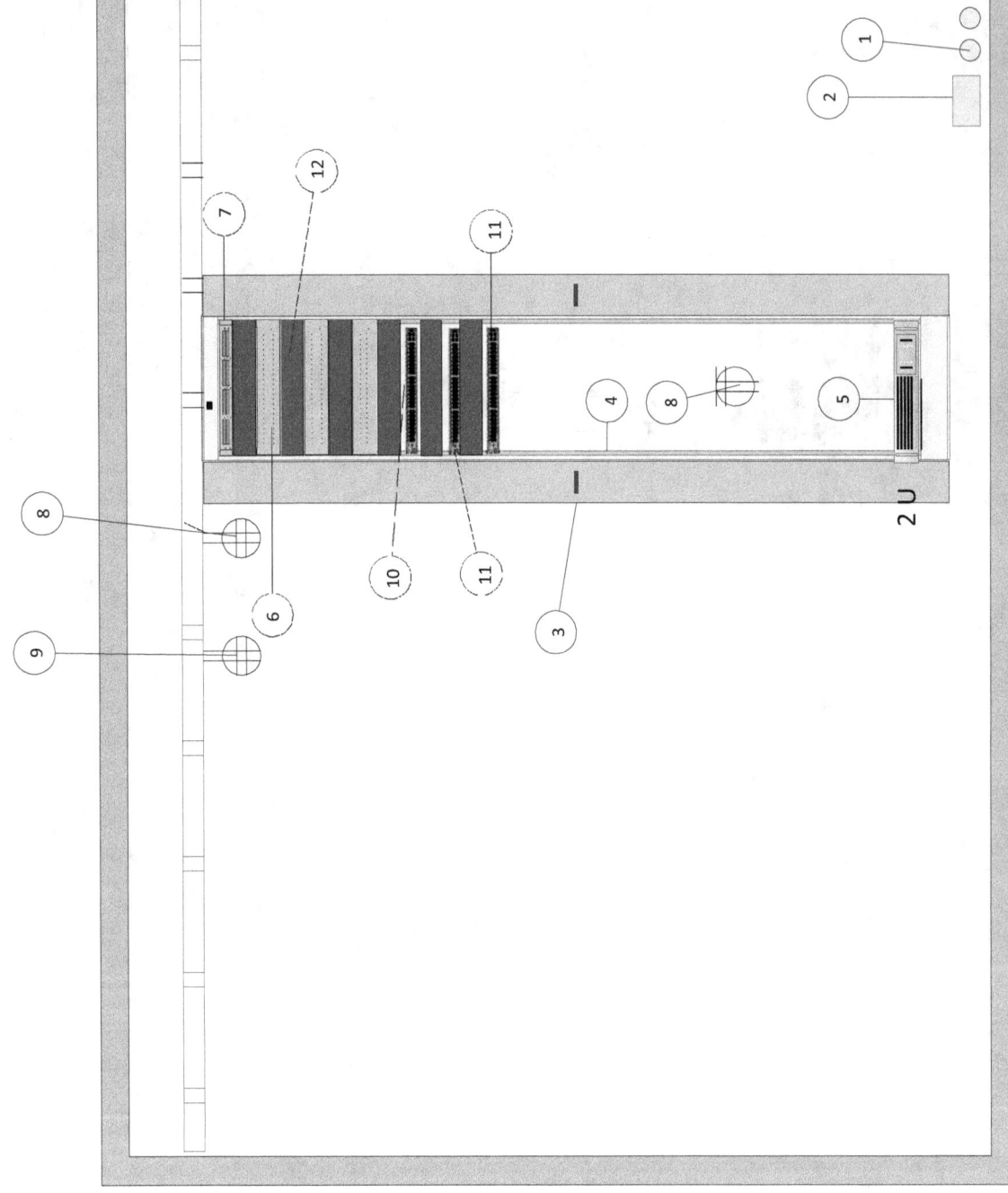

9' Original Drop Ceiling

1. Above Ceiling Core Bore 2-Each 4 ¼". Install 2-Each 4" threaded sleeves between exact location or changes. Fire stop &screw on cap on unused sleeves and label accordingly.

2. Ground bus bar (TMGB) Chatsworth assembly, ¼" X 4" X12" long, with insulators, mount 15" AFF, supply with grounding conductor per TIA-607-A. Termination of conductor must be a crimped two-hole long barrel lug.

3. Vertical Cable Managers Chatsworth Products 30092-508.

4. Chatsworth 19" x 7' CPI rack, P/N 55053-703 with vertical cable managers and doors & mounted 4-feet centered from wall to front of rack.

5. Liebert GXT3 On-Line UPS, 500-3000VA

6. Hubble UDX48E 48 PORT XCELERATOR JACK PANEL

7. Corning EDGE-01U-FP Pretium EDGE® Solutions Housing, Fixed, 1 rack units, holds 8 Pretium EDGE Solutions Modules or Panels (with field removable strain-relief plate).

8. Quad receptacles, 120 volt, 20 AMP, 1-duplex per circuit, powered from E-power panel. Show detail indicating exact placement in relation to each rack.

9. Quad receptacles, 120 volt, 30 AMP, 1-duplex per circuit, powered from E-power panel. Show detail indicating exact placement in relation to each rack.

10. Summit X480 2-Port 10 Gig 48-Port 1 Gig Ethernet, POE,

11. Summit X460 Series 48-Port Gig Ethernet.

12. Chatsworth Horizontal Cable Managers Chatsworth13930

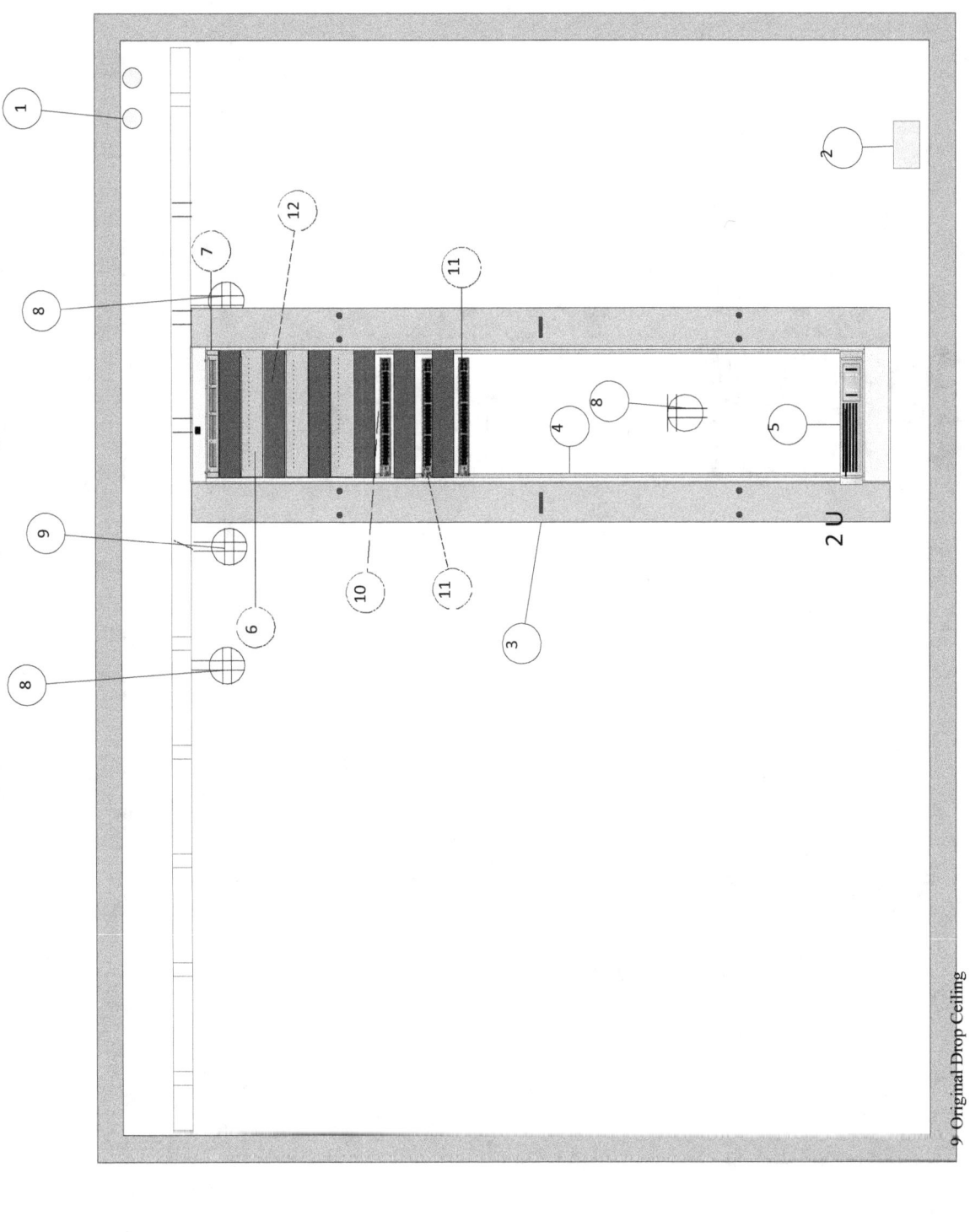

1. Above Ceiling Core Bore 2-Each 4 ¼. Install 2-Each 4" threaded sleeves between exact location or changes. Fire stop &screw on cap on unused sleeves and label accordingly.
2. Ground bus bar (TMGB) Chatsworth assembly, ¼" X 4" X12" long, with insulators, mount 15" AFF, supply with grounding conductor per TIA-607-A. Termination of conductor must be a crimped two-hole long barrel lug.
3. Vertical Cable Managers Chatsworth Products 30092-508.
4. Chatsworth 19" x 7' CPI rack, P/N 55053-703 with vertical cable managers and doors & mounted 4-feet centered from wall to front of rack.
5. Liebert GXT3 On-Line UPS, 500-3000VA
6. Hubble UDX48E 48 PORT XCELERATOR JACK PANEL
7. Corning EDGE-01U-FP Pretium EDGE® Solutions Housing, Fixed, 1 rack units, hold
8 Pretium EDGE Solutions Modules or Panels (with field removable strain-relief plate).
8. Quad receptacles, 120 volts, 20 AMP, 1-duplex per circuit, powered from E-power panel. Show detail indicating exact placement in relation to each rack.
9.Quad receptacles, 120 volts, 30 AMP, 1-duplex per circuit, powered from E-power panel. Show detail indicating exact placement in relation to each rack.
10. Summit X480 2-Port 10 Gig 48-Port 1 Gig Ethernet, POE,
11. Summit X460 Series 48-Port Gig Ethernet.
12. Chatsworth Horizontal Cable Managers Chatsworth 13930

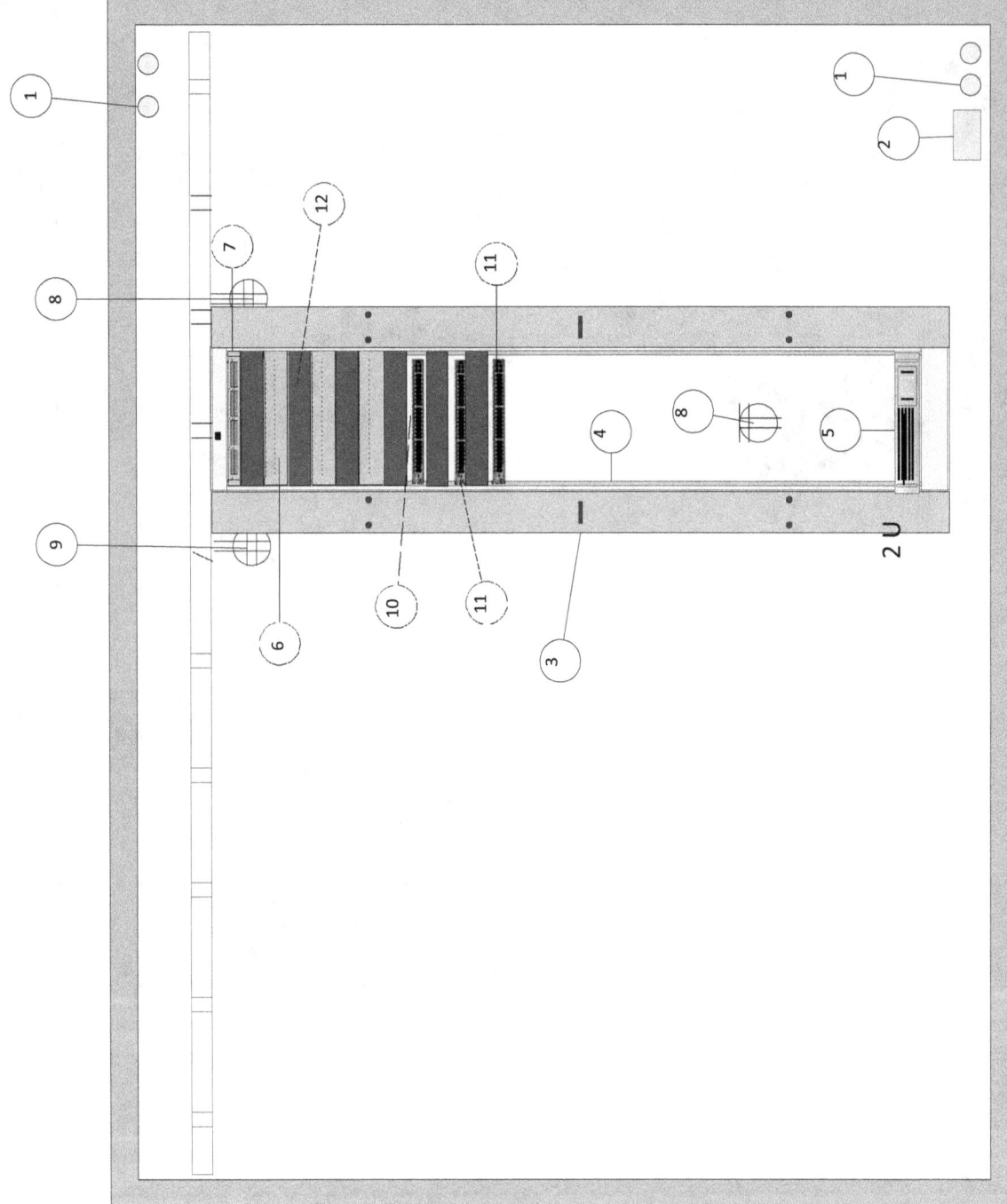

9' Original Drop Ceiling

1. Above Ceiling Core Bore 2-Each 4 ¼. Install 2-Each 4" threaded sleeves between exact location or changes. Fire stop &screw on cap on unused sleeves and label accordingly.
2. Ground bus bar (TMGB) Chatsworth assembly, ¼" X 4" X12" long, with insulators, mount 15" AFF, supply with grounding conductor per TIA-607-A. Termination of conductor must be a crimped two-hole long barrel lug.
3. Vertical Cable Managers Chatsworth Products 30092-508.
4. Chatsworth 19" x 7" CPI rack, P/N 55053-703 with vertical cable managers and doors & mounted 4-feet centered from wall to front of rack.
5. Liebert GXT3 On-Line UPS, 500-3000VA
6. Hubble UDX48E 48 PORT XCELERATOR JACK PANEL
7. Corning EDGE-01U-FP Pretium EDGE® Solutions Housing, Fixed, 1 rack units, hold
8 Pretium EDGE Solutions Modules or Panels (with field removable strain-relief plate).
8. Quad receptacles, 120 volts, 20 AMP, 1-duplex per circuit, powered from E-power panel. Show detail indicating exact placement in relation to each rack.
9.Quad receptacles, 120 volts, 30 AMP, 1-duplex per circuit, powered from E-power panel. Show detail indicating exact placement in relation to each rack.
10. Summit X480 2-Port 10 Gig 48-Port 1 Gig Ethernet, POE,
11. Summit X460 Series 48-Port Gig Ethernet.
12. Chatsworth Horizontal Cable Managers Chatsworth13930

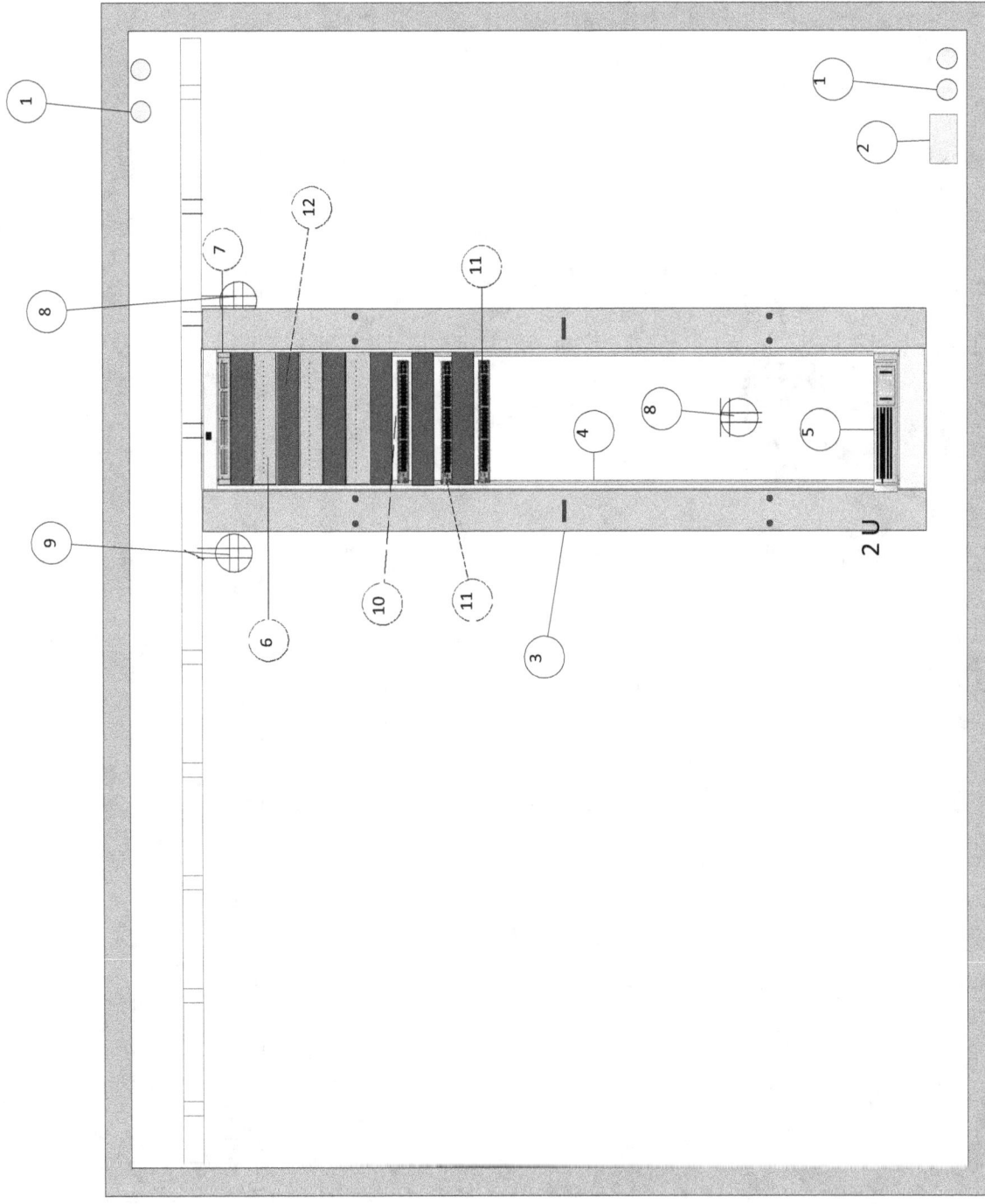

9 Original Drop Ceiling

1. Above Ceiling Core Bore 2-Each 4 ¼. Install 2-Each 4" threaded sleeves between exact location or changes. Fire stop &screw on cap on unused sleeves and label accordingly.
2. Ground bus bar (TMGB) Chatsworth assembly, ¼" X 4" X12" long, with insulators, mount 15" AFF, supply with grounding conductor per TIA-607-A. Termination of conductor must be a crimped two-hole long barrel lug.
3. Vertical Cable Managers Chatsworth Products 30092-508.
4. Chatsworth 19" x 7' CPI rack, P/N 55053-703 with vertical cable managers and doors & mounted 4-feet centered from wall to front of rack.
5. Liebert GXT3 On-Line UPS, 500-3000VA
6. Hubble UDX48E 48 PORT XCELERATOR JACK PANEL
7. Corning EDGE-01U-FP Pretium EDGE® Solutions Housing, Fixed, 1 rack units, hold 8 Pretium EDGE Solutions Modules or Panels (with field removable strain-relief plate).
8. Quad receptacles, 120 volts, 20 AMP, 1-duplex per circuit, powered from E-power panel. Show detail indicating exact placement in relation to each rack.
9.Quad receptacles, 120 volts, 30 AMP, 1-duplex per circuit, powered from E-power panel. Show detail indicating exact placement in relation to each rack.
10. Summit X480 2-Port 10 Gig 48-Port 1 Gig Ethernet, POE,
11. Summit X460 Series 48-Port Gig Ethernet.
12. Chatsworth Horizontal Cable Managers Chatsworth13930

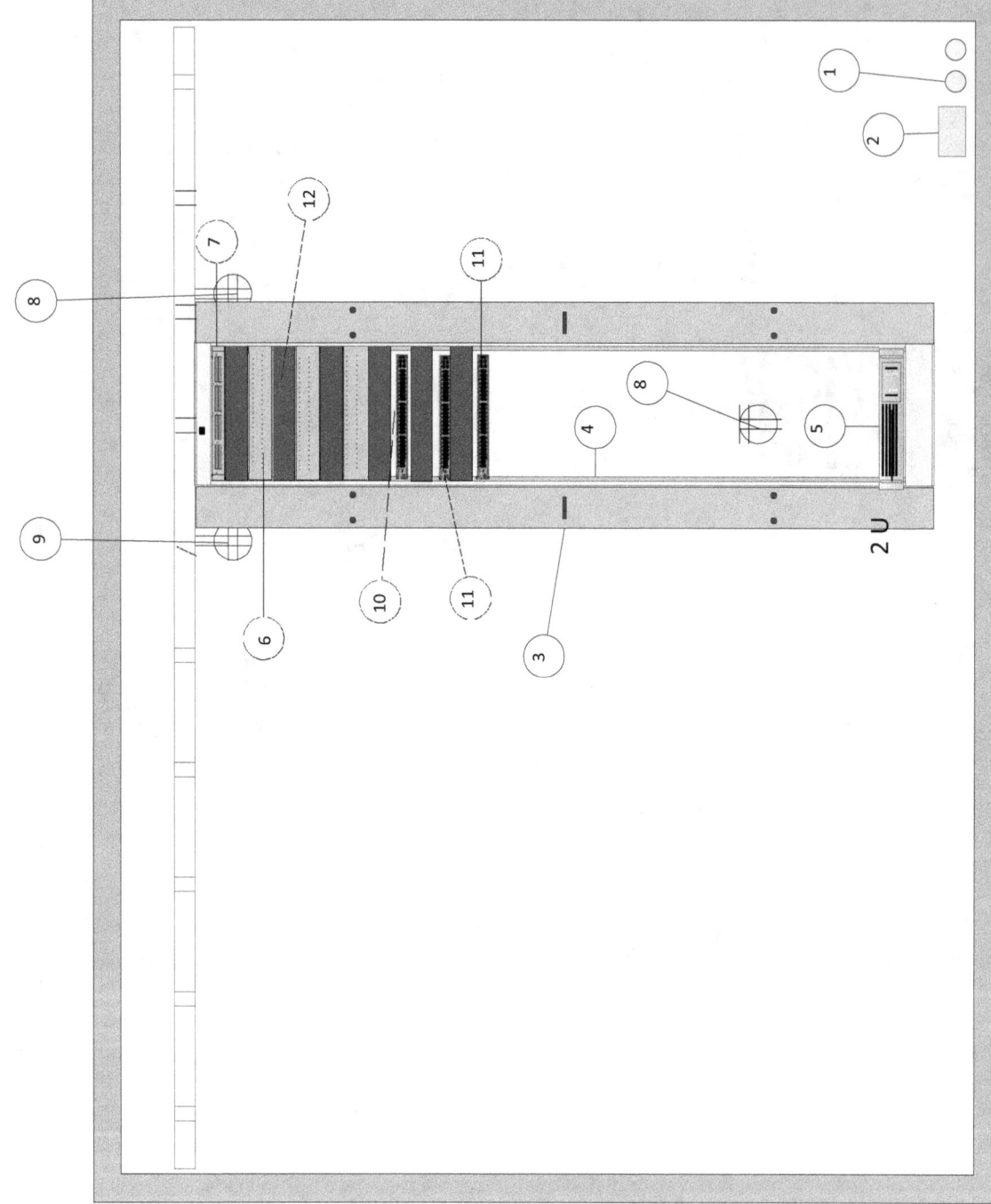

9' Original Drop Ceiling

1. Above Ceiling Core Bore 2-Each 4 ¼. Install 2-Each 4" threaded sleeves between exact location or changes. Fire stop &screw on cap on unused sleeves and label accordingly.
2. Ground bus bar (TMGB) Chatsworth assembly, ¼" X 4" X12" long, with insulators, mount 15" AFF, supply with grounding conductor per TIA-607-A. Termination of conductor must be a crimped two-hole long barrel lug.
3. Vertical Cable Managers Chatsworth Products 30092-508.
4. Chatsworth 19" x 7' CPI rack, P/N 55053-703 with vertical cable managers and doors & mounted 4-feet centered from wall to front of rack.
5. Liebert GXT3 On-Line UPS, 500-3000VA
6. Hubble UDX48E 48 PORT XCELERATOR JACK PANEL
7. Corning EDGE-01U-FP Pretium EDGE® Solutions Housing, Fixed, 1 rack units, hold
8 Pretium EDGE Solutions Modules or Panels (with field removable strain-relief plate).
8. Quad receptacles, 120 volts, 20 AMP, 1-duplex per circuit, powered from E-power panel. Show detail indicating exact placement in relation to each rack.
9.Quad receptacles, 120 volts, 30 AMP, 1-duplex per circuit, powered from E-power panel. Show detail indicating exact placement in relation to each rack.
10. Summit X480 2-Port 10 Gig 48-Port 1 Gig Ethernet, POE,
11. Summit X460 Series 48-Port Gig Ethernet.
12. Chatsworth Horizontal Cable Managers Chatsworth13930

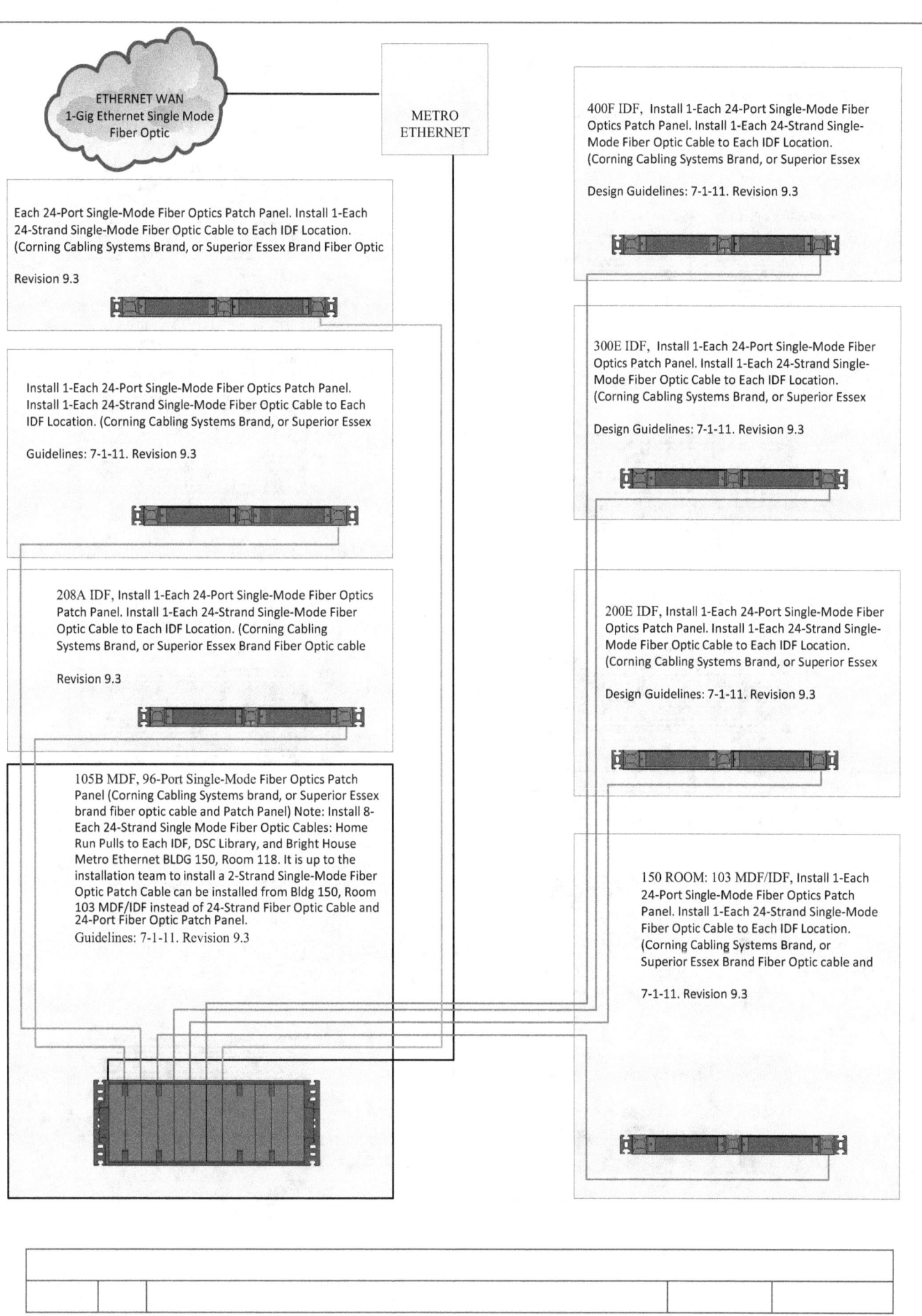

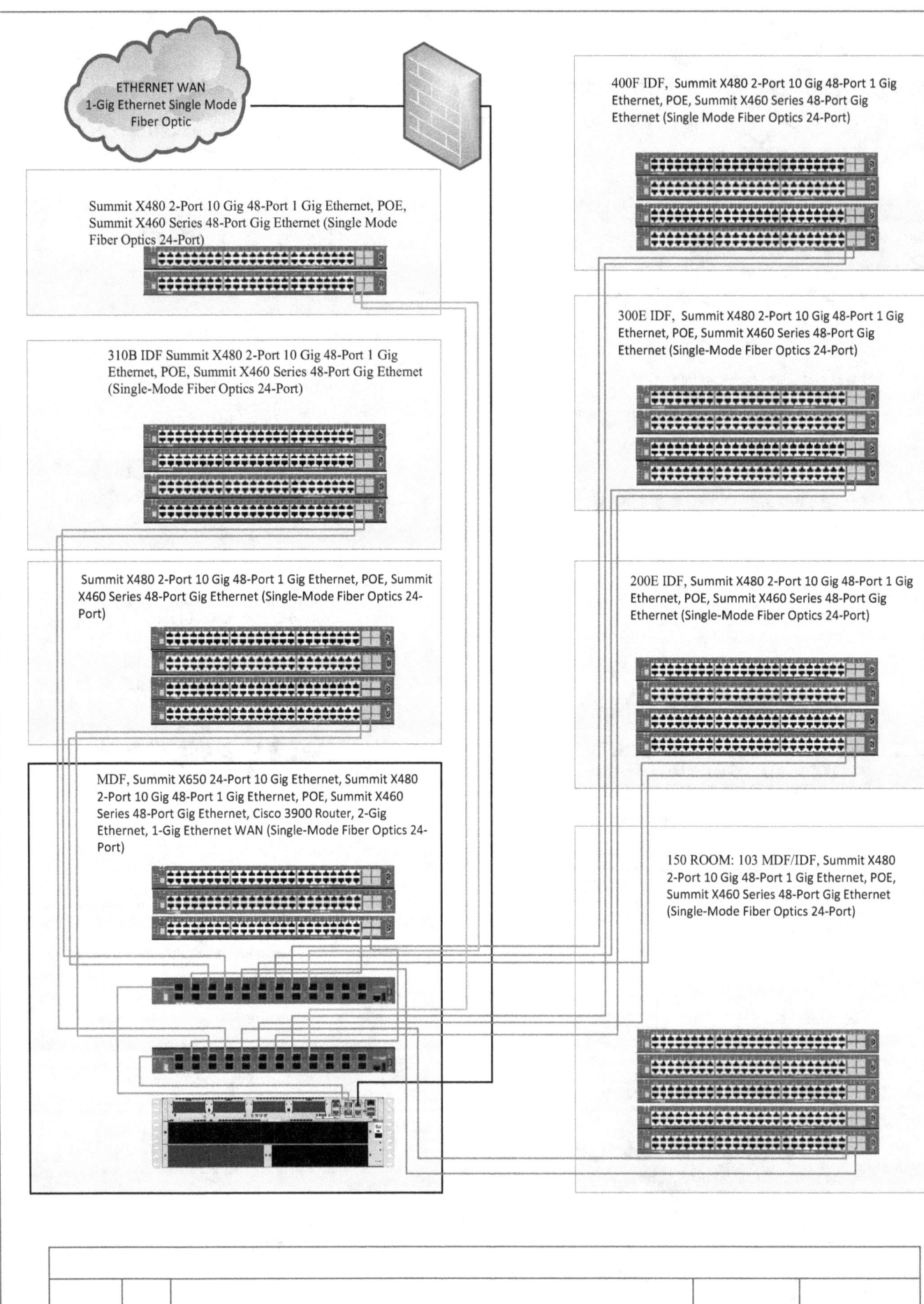

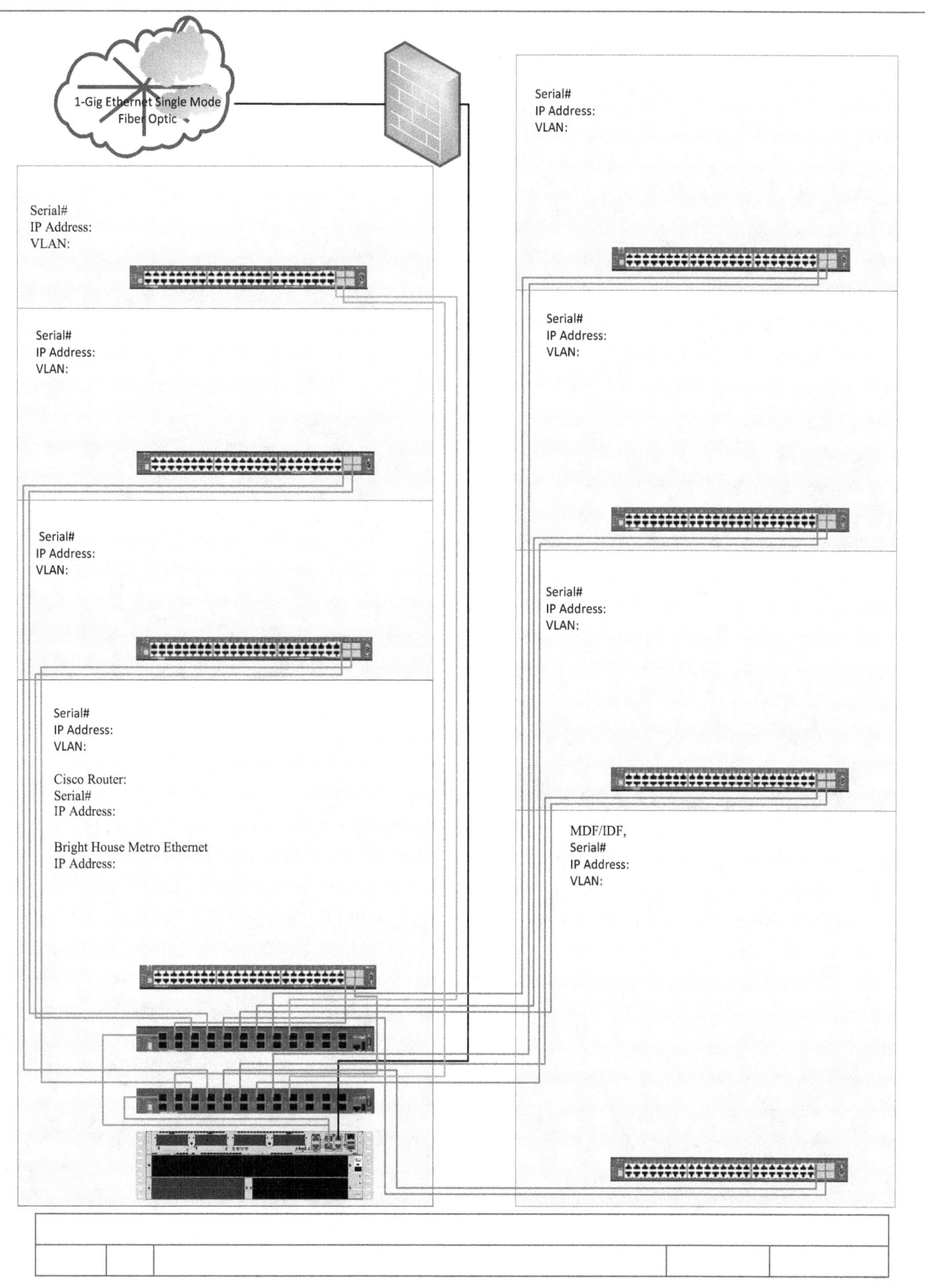

www.ingramcontent.com/pod-product-compliance
Lightning Source LLC
Chambersburg PA
CBHW080910220526

45466CB00011BA/3531